MAJESTIC LIGHTS

The Aurora in Science, History, and the Arts

ROBERT H. EATHER

American Geophysical Union
Washington, D. C.

MAJESTIC LIGHTS

The Aurora in Science, History, and the Arts

Library of Congress Catalog Number: 79-56387
ISBN: 0-87590-215-4

Permission to reprint from the following publishers is gratefully acknowledged.

Reprinted by permission of Dodd, Mead and Company, Inc. from *The Collected Poems
of Robert Service.* Copyright © 1917 by Dodd, Mead and Company, Inc. Copyright
renewed in 1945 by Robert Service.

Reprinted by permission of D. Reidel Publishing Company from *The Physicist's Concept of Nature,* edited by J. Mehra, from M. Eigen, "The Origin of Biological Information," pp. 594–632. Copyright © 1973 by D. Reidel Publishing Company.

Reprinted by permission of Doubleday and Company from *Knowledge and Wonder* by V. Weisskopf. Copyright © 1967 by Doubleday and Company.

Reprinted by permission of John Murray (Publishers) Ltd. from *Scott's Last Expedition* by R. F. Scott.

The Eskimos by E.M. Weyer. Copyright © 1932 by Yale University Press. Reprinted with permission.

North/Nord, Gromnica. Copyright © by the Department of Indian and Northern Affairs, Government of Canada. Reprinted by permission.

Reprinted by permission of the Hakluyt Society. Bellinghausen, T., *The Voyage of Captain Bellinghausen to the Antarctic Seas, 1891–1921.* Copyright © 1945, The Hakluyt Society, London.

Reprinted by permission of Pergamon Press Ltd. Robertson, T. A., *The Northern Lights* in *Keoeeit — The Aurora* by W. Petrie.

Gartlein, C. W., "Unlocking the Secrets of the Northern Lights," *National Geographic Magazine.* Copyright © 1947, National Geographic Society. Reprinted with permission.

Reprinted by permission of Hughes Massie Limited, London. Amundsen, R., *The South Pole.*

Reprinted from *The Collected Poems of Wallace Stevens* by permission of Alfred A. Knopf, Inc. Stevens, W., "The Auroras of Autumn, 1949." Copyright © 1948 by Wallace Stevens.

Reprinted by permission of Oxford University Press, Inc. from *Science and English Poetry: A Historical Sketch, 1590–1950* by John Nash Douglas Bush. Copyright © 1950 by Indiana University, Bloomington, Indiana; renewed in 1977 by John Nash Douglas Bush.

Reprinted by permission of Weidenfeld (Publishers) Limited, London. Solzhenitsyn, A. I., *Matryana's House,* in *Halfway to the Moon — New Writings From Russia.*

Reprinted by permission of the American Scandanavian Foundation. Benediktsson, E., "Northern Lights" in *Icelandic Poems and Stories.*

Reprinted by permission of John Murray (Publishers) Ltd. Jacob, V., *Northern Lights and Other Poems.*

Reprinted by permission of the Trustees of Amherst College from *The Poems of Emily Dickinson,* edited by Thomas H. Johnson, Cambridge, Massachusetts: The Belknap Press of Harvard University Press. Copyright © 1951 and 1955 by the President and Fellows of Harvard College.

Reprinted by permission of G. P. Putnam's Sons from *Alone* by Richard E. Byrd. Copyright © 1938 by Richard E. Byrd. Renewed © 1966 by Marie A. Byrd.

Hopkins, G. M., "Poems and Prose of Gerald Manley Hopkins". Reprinted from *The Journals and Papers of Gerald Manley Hopkins* edited by Humphrey House and Graham Storey (2nd ed.) 1959. By permission of Oxford University Press.

Used by permission of Atheneum Publishers. Aiken, C., "Tigermine." Poems copyright © 1977 by Mary Hoover Aiken. Illustrations copyright © 1977 by John Vernon Lord.

Reprinted by permission of Harcourt, Brace, and Jovanovich, Inc. Eliot, T. S., "The Love Song of J. Alfred Prufrock" in *Collected Poems 1909–1932.*

Reprinted by permission of Simon and Schuster, Inc. Bronowski, J., *Science and Human Values.* Copyright © 1956 by J. Bronowski.

Reprinted by permission of Martin Secker and Warburg Ltd. and Alfred A. Knopf, Inc. Gabor, D., *Inventing the Future.* Copyright © 1963.

Reprinted by permission of New York University. Heyl, P. R., *Physics.*

Reprinted by permission of the University of Chicago Press. Szent-Gyorgi, A., "Perspectives in Biology and Medicine" in *The Harvest of a Quiet Eye.* Copyright © 1971 by A. Szent-Gyorgi.

Reprinted by permission of McGraw-Hill Book Company from *Assault on the Unknown* by Walter Sullivan. Copyright © 1968 by Walter Sullivan.

Aurora, the Roman goddess of dawn, identified with the Greek goddess *Eos.* This painting from an ancient Greek vase shows Homer's rosy fingered goddess borne on wings of air dispensing the dew of the morning from two large urns.

The most beautiful experience we can have is the
mysterious. It is the fundamental emotion which
stands at the cradle of true art and true science.
Whoever does not know it and can no longer wonder,
no longer marvel, is as good as dead,
and his eyes are dimmed.

Albert Einstein
The World as I See It
in *Living Philosophies,* 1931

Some say that the Northern Lights
are the glare of the Arctic ice and snow;
And some say that it's electricity,
and nobody seems to know.

Robert W. Service
Ballad of the Northern Lights, 1909

Table of Contents

Preface ix
1. Modern Aurora Watching 1
2. The First 2000 Years 33
3. The Beginning of Scientific Enquiry 47
4. Severnoe Sijanie 65
5. A Century of Observations 75
6. The Aurora in Colonial America 93
7. Legends and Folklore 103
8. Aurora Bright, Rain Tonight? 117
9. The Emergence of Norwegian Auroral Science 129
10. Aurora Australis 141
11. Auroral Audibility .153
12. The Contribution of the Spectroscopists 163
13. The International Geophysical Year 171
14. The Aurora in Poetry and Literature 185
15. The Magnetosphere .215
16. Man-Made Aurora .231
17. What Now? 243
18. Auroral Photography 251
19. The Aurora and Me .259
Appendix. Names for the Aurora 280
Photograph Gallery 283
Acknowledgments 291
Bibliography 299
Name Index 317
Subject Index 319

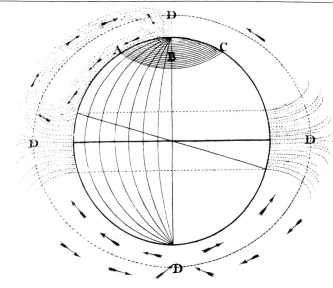

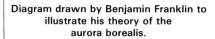

Diagram drawn by Benjamin Franklin to illustrate his theory of the aurora borealis.

The Arrows represent the general Currents of the Air.
A.B.C. the great Cake of Ice & Snow in the Polar Regions.
D.D.D.D. the Medium Height of the Atmosphere.
The Representation is made only for one Quarter and one
Meridian of the Globe; but is to be understood the same
for all the rest.

Benjamin Franklin 1706–1790
Statesman, scientist, inventor,
and author

B.FRANKLIN, L.L.D. F.R.S.

Born at Boston in New England, Jan 17th 1706.

NON SORDIDUS AUCTOR NATURÆ VERIQUE.

Preface

> The atmosphere being heavier in the polar regions, than in the
> equatorial, will there be lower; as well as from that cause, as from
> the smaller effect of the centrifugal force: consequently the distance of
> the vacuum above the atmosphere will be less at the poles, than
> elsewhere; . . . May not then the great quantity of electricity, brought
> into the polar regions by clouds, which are condensed there, and fall
> in snow, which electricity would enter the earth, but cannot penetrate
> the ice; may it not, I say, break through that low atmosphere, and run along
> in the vacuum over the air and towards the equator; diverging as the
> degrees of longitude enlarge; strongly visible where densest, and becoming
> less visible as it more diverges; till it finds a passage to the earth in
> more temperate climates, or is mingled with their upper air.
>
> If such an operation of nature were really performed would it not give
> all the appearances of an *Aurora Borealis?*

So wrote Benjamin Franklin (176), L.L.D., F.R.S., in a paper read to the
Royal Academy of Sciences at Paris at the meeting held immediately
after Easter in 1779. Franklin is but one famous name among men of
science who has puzzled over the aurora since science began. A partial
list reads almost like a *Who's Who* of science up until our present century, and it includes Aristotle, Seneca, Kepler, Galileo, Gassendi, Halley,
Euler, Descartes, Celsius, Cavendish, Dalton, Volta, Gauss, Humbolt,
and Angström.

To trace the story of the aurora through history is to trace the development of man from a creature of ignorance and superstition, through his
renaissance of art and learning, to an analytical disciple of science and
technology. But fortunately, man's transformation is not complete:
Nature has contrived to clothe all objects of our scientific investigation,
from the microscopic to the cosmic, in an aura of beauty and surprise.
The aurora is perhaps the most spectacular of nature's contrivances to
preserve the soul of the scientist.

1
Modern Aurora Watching

This gorgeous apparatus
This display
This ostentation of creative power
This theatre
What eye can take it in . . .?

Edward Young
The Complaint, or Night Thoughts,
Night Ninth, 1744

Most readers of this book will have never seen the northern lights, though everyone will have heard of the phenomenon and have at least some vague idea that it is "lights in the sky that appear at the north pole." So before we embark on the fascinating story of the 2500 years of development of our understanding of these lights, this first chapter presents a variety of photographs to show the beauty and variety of the aurora, briefly describes what it is and how it happens, and discusses the chances that most people might have of ever witnessing a grand display.

The variety of shapes, colors, structures, and movements of the aurora are infinite. Like snowflakes, no two are ever quite the same. When faced with such a bewildering plethora, scientists try to feel more in control of the situation by classifying auroras into types and subtypes. The first such auroral classification system was devised by the famous Norwegian auroral physicist Carl Störmer (see page 135) and published in the *Photographic Atlas of Auroral Forms* in 1930 (446). His system was revised in 1963 when a new *International Aurora Atlas* was published (245).

Rather than go into the details of that classification system, we have abstracted pertinent information that would give an amateur observer the basic vocabulary that would be needed to describe adequately an aurora to a scientist. The accompanying pictures give examples of the various classifications.

Five forms are defined as fundamental in the identification and reporting of auroras: *arcs*, which appear as simple slightly curving arcs of light with smooth lower borders; *bands*, which have continuous but irregular lower borders characterized by kinks or folds; *patches*, which are isolated small regions of luminosity, often resembling patches of cloud; *veil*, which describes an extensive, uniform luminosity which covers a large fraction of the sky; and *rays*, which are shafts of luminosity inclined to the vertical (in the direction of the earth's magnetic field). These five forms commonly exhibit one of three types of structures: *homogeneous*, or the lack of internal structure, so that the brightness is uniform; *striated*, where irregular fine striations, or filaments, are seen; or *rayed*, where rays appear within forms, such as rayed arcs or bands.

A special form of the rayed band is present when the band exhibits long rays and has an overall folded structure. Such auroras resemble draperies or curtains and are often referred to by those terms. An even more spectacular form of the rayed form is the *corona*; this is a perspective effect when a rayed form is overhead and all rays appear to converge to a point (see page 15). The point of convergence (radiant point) is the local direction of the earth's magnetic field.

Auroras are also classified according to their temporal behavior: *quiet*, where the form is uniform in intensity over long periods; *pulsating*, where the brightness increases and decreases in a quasiperiodic fashion. Some special cases of the pulsating aurora are flickering aurora, where the brightness flickers at a high frequency (say, 5–10 times per second), and flaming aurora, where bursts of luminosity appear at the base of the form, rapidly move vertically up the form, and disappear at the top.

Any auroral type may have a large range of brightness, classified between 0 (subvisual) and 4 (very bright). These brightness classifications are explained in the section on photographing the aurora (see page 257). Professional observers also utilize more subtypes depending on movements, intensity fluctuations, and color.

Now that we can describe the aurora, what causes it? This, of course, is the topic of this book, but let us begin with my usual 1-minute cocktail party explanation:

The sun is a very hot gaseous body, so hot that it continually "boils off" atomic particles—protons and electrons. These particles stream out from the sun in all directions into interplanetary space and constitute the so-called "solar wind." After a couple of days a small fraction of these particles reaches the vicinity of earth. Their first indication that anything unusual is in the offing is when they encounter the earth's magnetic field, which stretches tens of thousands of miles out into space. Magnetic fields affect the trajectories of electrically charged particles, and in this case, the shape of the earth's magnetic field acts to guide some of these solar wind particles to two oval-shaped regions around the north and south poles. These ovals are located some 23° latitude from the poles, and this is the region where auroras occur every night of the year. As these solar wind protons and electrons are being guided down to lower altitudes of the earth's atmosphere, they are also accelerated to higher velocities. The fast particles then collide with the earth's upper atmosphere—the oxygen and nitrogen. A process then occurs which is analogous to the operation of a neon light tube: in the light tube, fast electrons collide with neon gas and give some of their energy to the gas, which is then radiated in the form of light. The same thing happens in the earth's atmosphere (at a height of about 60 miles), except here it is the oxygen and nitrogen that emit light. This is the light of the aurora, and its color depends on the height. The faster the precipitating particles, the deeper they penetrate into the atmosphere. But the composition of the atmosphere—the relative amounts of oxygen and nitrogen atoms and molecules—varies with height, and each gas gives out its own particular color when bombarded by these fast particles. Consequently, certain colors originate preferentially from certain heights in the sky, and the range in velocity of the incoming particles results in the wonderful variety of color of the aurora.

So much for simple explanations—not all of the above is quite accurate, and large parts of the story have been omitted—though this short piece of pedagogy is certainly closer to the truth than are the inaccuracies that continue to be perpetrated by many modern encyclopedias and school textbooks. A satisfactory explanation of the aurora is a recent development and has only been possible within the last 20 years. In fact, auroral science has probably progressed more in the past 20 years than in the previous 2000, and these advances in auroral research have closely paralleled and been linked to modern technological advances.

The auroral photographs in this chapter show something of the beauty of the aurora, though anyone who has ever seen a good display in the Arctic knows that it cannot be captured adequately on film. What chance then does a resident of more southern climes have of seeing the northern lights, short of buying an air ticket to Alaska? The map on page 4 shows the percentage of nights per year on which auroral displays are expected at various latitudes for the northern hemisphere. Of course, this is a statistical result and may not hold for any given year, but it does give an idea of the probability of displays. On the other hand, it does not take into account that it might be cloudy and does not guarantee a bright and spectacular display. This information for various major cities is given in the tabulation to the right.

Examination of the tabulation shows that for most cities the chances of seeing an aurora are depressingly low. And with the high ambient light levels of cities, an aurora would have to be very bright indeed to even be noticed. However, there are a number of steps an interested person may take to maximize the chances of seeing an aurora when it does occur.

Percentage of Nights on Which Aurora Is Expected	
City	Percentage
Northern Hemisphere	
Barrow, Alaska	100
Churchill, Canada	100
Fairbanks, Alaska	90
Anchorage, Alaska	30
Winnipeg, Canada	20
Calgary, Canada	18
Montreal, Canada	10
Bangor, Maine	9
Buffalo, New York	6
Minneapolis, Minnesota	6
Boston, Massachusetts	5
Seattle, Washington	5
New York, New York	4
Chicago, Illinois	4
Washington, D. C.	3
Denver, Colorado	3
Atlanta, Georgia	1.5
San Francisco, California	1.5
Los Angeles, California	0.5
Houston, Texas	0.5
Mexico City, Mexico	~0.05
Tromso, Norway	90
Kiruna, Sweden	80
Oslo, Norway	10
Edinburgh, Scotland	8
Moscow, USSR	3
London, England,	2.5
Paris, France	1.5
Munich, Germany	1
Madrid, Spain	0.2
Rome, Italy	0.1
Tokyo, Japan	~0.01
Southern Hemisphere	
Melbourne, Australia	3
Sydney, Australia	1
Auckland, New Zealand	1
Capetown, South Africa	0.5
Rio de Janiero, Brazil	~0.01
Buenos Aires, Argentina	~0.01

3

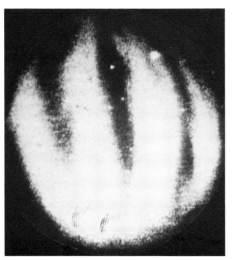

North-south aligned equatorial depletion regions seen in red airglow emission.

Airglow

When we view auroral forms, they are superimposed on an overall background luminosity in the night sky. We are aware of the localized sources of light in a moonless night sky (the stars and planets, the zodiacal light, and gegenschein), but in addition to the astronomical sources there is an overall uniform luminosity originating from the earth's own atmosphere. We are not normally aware of this *airglow* (or *nightglow* as it is sometimes called) because it is so uniform. It is the combination of astronomical and airglow sources that allows us to see the silhouette of an object held against the "dark" sky on a clear moonless night.

The airglow contributes some 25–50% to the total light of the night sky (at locations well away from the auroral regions). It results from photochemical reactions between oxygen and nitrogen atoms and molecules and hydroxyl molecules at a height between 100 and 300 kilometers. Solar radiation energy breaks molecules apart during the day, and it is their recombination during the night, accompanied by the emission of light, that generates the airglow. Most of the spectral emissions of the airglow are also seen in the aurora, but the aurora has many additional emissions and, of course, is very much brighter.

During magnetic storms the airglow at mid-latitudes is modified by heating effects, producing "red arcs" (also called M-arcs and SAR-arcs). These are broad diffuse arcs of light that align along geomagnetic parallels between about 45°-55° and locate at a height of 300-400 kilometers. They are quite bright in red oxygen emission but are subvisual because the human eye's sensitivity at that wavelength is only about 5% of its peak sensitivity in the yellow region of the spectrum.

Even closer to the geomagnetic equator, within about 5°, another airglow phenomenon is observed. It is again subvisual, is high in the atmosphere and seen in the red oxygen emission. North-south aligned airglow depletion regions are recorded by instruments, and the regions drift in easterly and westerly directions. When recorded by sensitive image-intensified television systems and then played back at high speeds, these movements bear an uncanny resemblance to the polar aurora. These depletions are caused by height changes of the upper ionosphere as "bubbles" of plasma move upward in this region.

Map of northern hemisphere showing percentage of nights on which one can expect to see an aurora (weather conditions permitting).

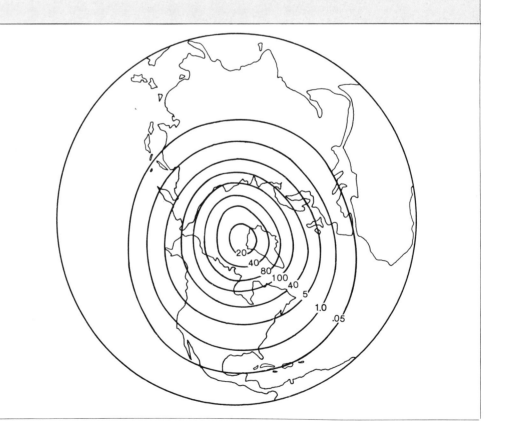

Auroras are seen at lower latitudes following unusual activity on the surface of the sun. At these times many more particles are thrown out into the solar wind and stream toward the earth. This journey takes 2–3 days. Large solar events usually rate a small news story in the daily newspapers, so that there is a 2-day warning when auroras might be seen. From 2 to 4 days after the solar event, a regular check should be kept on the northern sky (southern sky in the southern hemisphere), especially between about 9 P.M. and midnight.

Rather than relying on noticing a small article in the newspaper, one can take more positive steps by utilizing the services of the Space Environmental Services Center in Boulder, Colorado. This is the only facility in the world dedicated to real time monitoring and prediction of the space environment — the energetic weather blown toward us in the solar wind. Various observatories around the world continually monitor activity on the sun and relay the information to Boulder; in addition, satellites make direct measurements of the solar wind and transmit the information back to Boulder.

From this steady flow of incoming data, forecasts are prepared for up to 27 days (the sun's rotation period) of solar activity and its probable geophysical effects, including the aurora. Detailed forecasts are sent out by teletype to interested organizations, and summaries are always available to members of the general public. This information is broadcast as part of the WWV time signals at 5, 10, and 15 MHz and received around the world on shortwave radio equipment such as widely available "Time Cubes." The 40-second message is broadcast at 18 minutes past each hour; it summarizes solar and geophysical activity for the previous 24 hours and gives predictions for the next 24 hours.

More convenient for residents of the United States and Canada is a tape-recorded message that may be heard by telephoning the center in Boulder (24 hours a day); inquiries may also be made directly to the duty forecaster (7 A.M. to midnight, 7 days a week). The phone numbers and format of the recorded message are given on the next page.

Finally, if you are traveling or vacationing in northern areas and are interested in trying to see or photograph the aurora, you may write to the Boulder center and request a copy of the current weekly publication *Preliminary Report and Forecast of Solar and Geophysical Activity*, which summarizes the previous week and predicts for the next 7- and 27-day periods.

Even with these hints and the forecasting services that are available, there is still no guarantee that nature will cooperate. The northern lights have the uncanny habit of appearing unannounced and unexpectedly. There will be a great aurora in the sky above your head sometime during your lifetime. I wish you luck that you happen to look up and be privileged to see nature's spectacular light show!

The format below is used for broadcast messages.

Space Environment Services Center

40-second message broadcast on WWV at 18 minutes past each hour. Tape-recorded message by dialing (303) 499-8129. Duty forecaster inquiries by dialing 303-449-1000 ext. 3171.

Message Format

A.   Solar terrestrial conditions for_____follow:

                                                            (month - day(s))

(Always includes first two)

  (1)   Solar flux_____

  (2)   A-index_____

  (3)   Solar activity was   (very low) (low) (moderate) (high) (very high)

  (4)   Geomagnetic field was   (quiet)   (unsettled)   (active)

       or A (minor) (major) geomagnetic storm (began at _____ UT)

                                                      (day - hour)

                       (is in progress)

                       (ended at _____ UT)

                                                      (day - hour)

  (5)   Major solar flare at _____

                           (coordinates)

  (6)   Proton flare at _____

                         (coordinates)

  (7)   Proton event on satellite (began at _____ UT)

                                                (day - hour)

                       (is in progress)

                       (ended at _____ UT)

                                                (day - hour)

  (8)   Polar cap absorption event (began at _____ UT)

                                                (day - hour)

                       (is in progress)

                       (ended at _____ UT)

                                                (day - hour)

B.   The forecast for _____ follows:

                                   (month - day)

(Includes any of the following that pertain)

  (1)   Solar activity will be   (very low) (low) (moderate) (high) (very high)

  (2)   Geomagnetic field will be   (quiet)   (unsettled)   (active)

       or Geomagnetic storm expected _____

                                              (date(s))

  (3)   Proton flares expected _____

                                         (coordinates)

  (4)   Stratwarm alert provided by National Weather Service

          _____

          _____

  This information is provided by the NOAA Space Environment Laboratory on behalf of the International Ursigram and World Days Service.

For descriptive leaflets, write to: Space Environment Services Center
                                    NOAA
                                    Boulder, Colorado 80303
                                    U.S.A.

The Sun has been observed by man for only a very small portion of its 4.6-billion-year life. Many ancient civilizations believed the Sun contained magical properties and was the giver of life; they worshiped the gaseous ball as a god. Other early cultures made accurate astronomical observations of the Sun, as seen in the neolithic cromlech of Stonehenge and in the Aztec solar calendar. When Galileo turned the newly invented telescope toward the sky less than 400 years ago to record sunspots and other phenomena, he marked the way for enormous advancements in man's continued observations of the Sun.

The Sun goes through cycles of high and low activity that recur at approximately 11-year intervals.

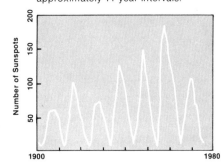

The console displays some of the data that forecasters use to predict solar and magnetic activity and to issue world-wide alerts of conditions that could jeopardize scientific projects.

The SPACE ENVIRONMENT SERVICES CENTER (SESC) observes solar activity extensively. The Center is interested in the complex and dynamic Earth-Sun relationship and especially in extremes in solar phenomena. In order to describe, understand, and predict this relationship, SESC acquires real-time solar data, makes forecasts of solar activity, and conducts research to improve observations and forecasting techniques.

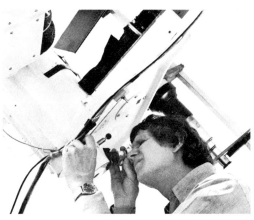

SESC monitors solar activity with an optical telescope and other sensing equipment in Boulder, and at cooperating observatories around the world. Round-the-clock reports on solar events are also received from orbiting satellites that measure the flow of solar particles and changes in the geomagnetic field.

Solar activity is manifested in plages, flares, sunspots, and filaments on the Sun's surface. During peaks of activity, the number of these active areas greatly increases.

During a solar flare, charged particles bombard the Earth's atmosphere at the Poles. One of the most beautiful effects of this solar storm is the aurora borealis, or northern lights.

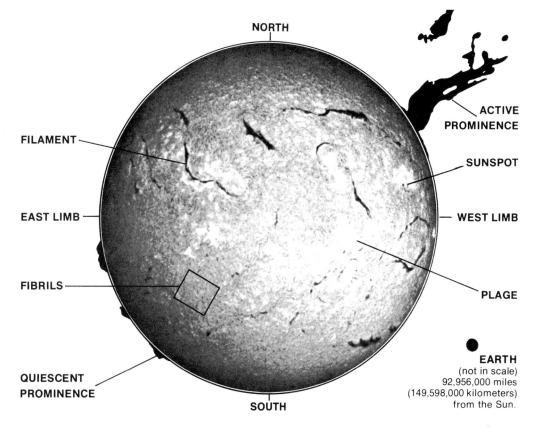

NORTH

FILAMENT

EAST LIMB

FIBRILS

QUIESCENT PROMINENCE

SOUTH

ACTIVE PROMINENCE

SUNSPOT

WEST LIMB

PLAGE

EARTH
(not in scale)
92,956,000 miles
(149,598,000 kilometers)
from the Sun.

The Space Environment Services Center Boulder, Colorado

By Earth standards, the Sun is tremendous, yet it is only an average size star with an average stellar magnitude (brightness). At its middle latitudes, this ball of helium and hydrogen rotates from east to west, with respect to Earth, once every 27 days; polar rotation is slower, taking approximately 29 days. Temperatures in this solar nuclear furnace range from about 6000 K at the surface to millions of degrees in its interior.

(The solar disk displayed here is a NOAA photo taken with a hydrogen-alpha filter. Prominences are graphic representations.)

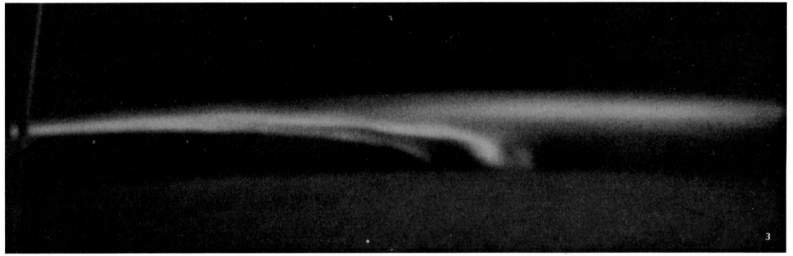

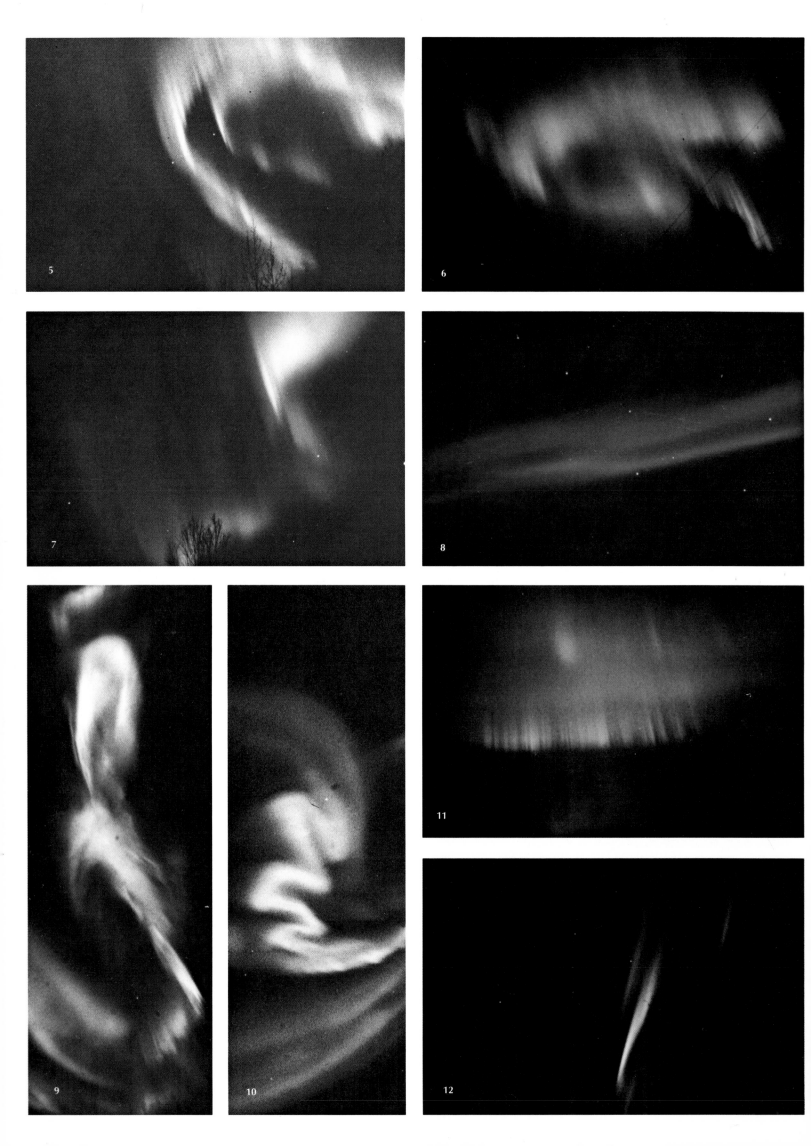

21

24

22

23

25

26

27

28

29

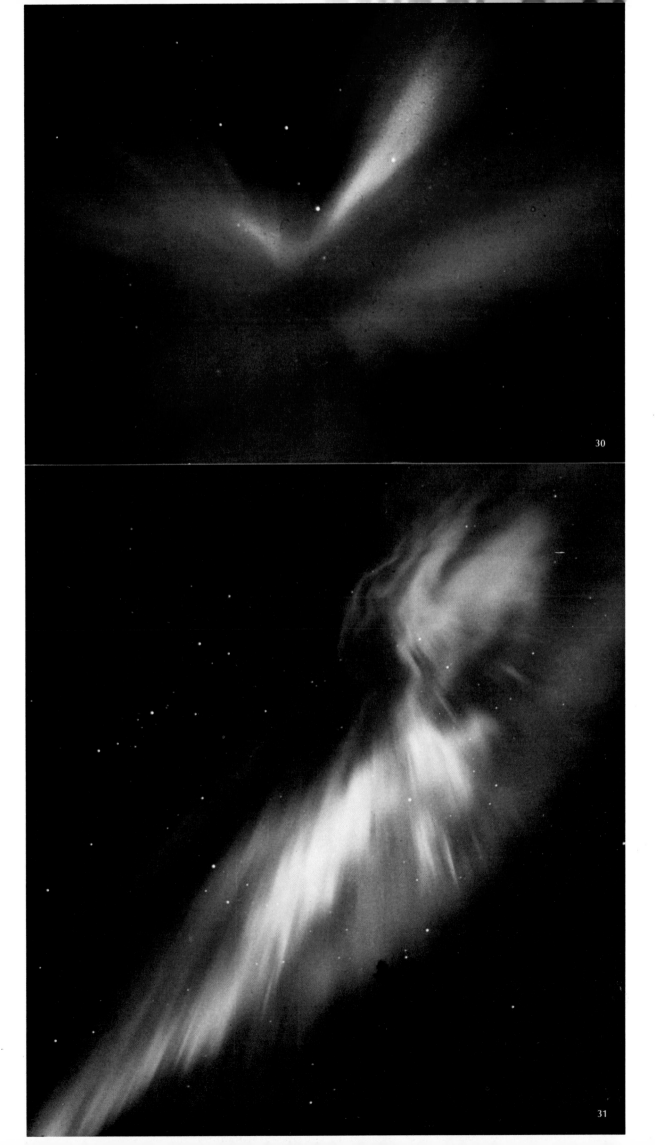

30

31

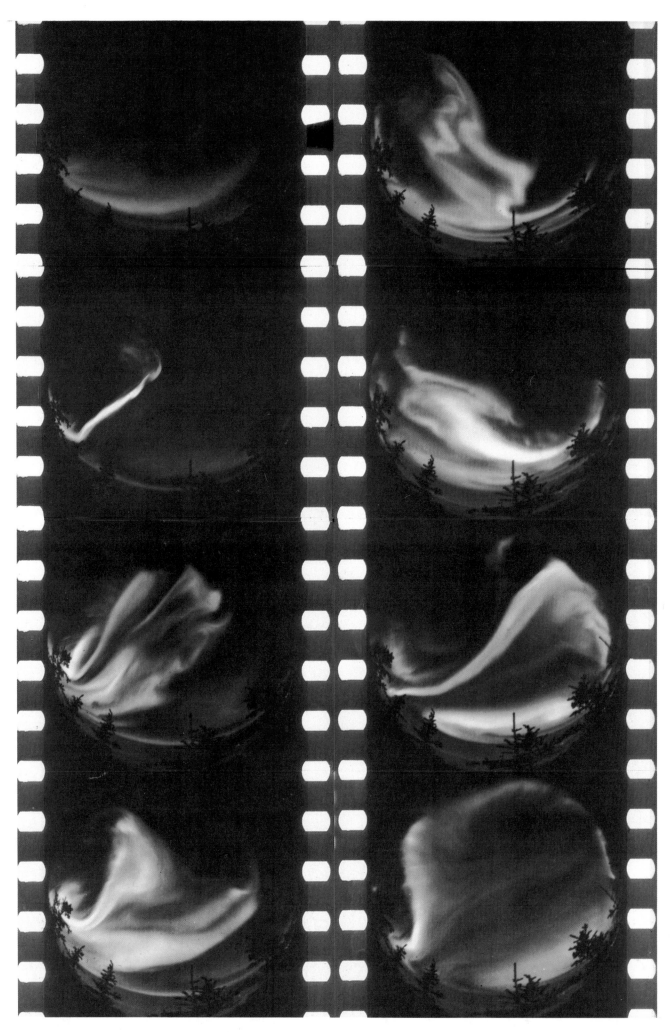

32

34

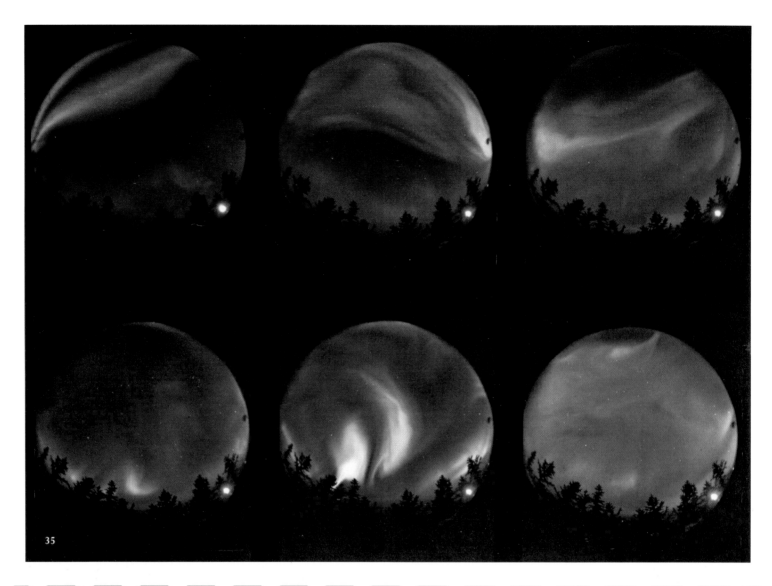

35

36

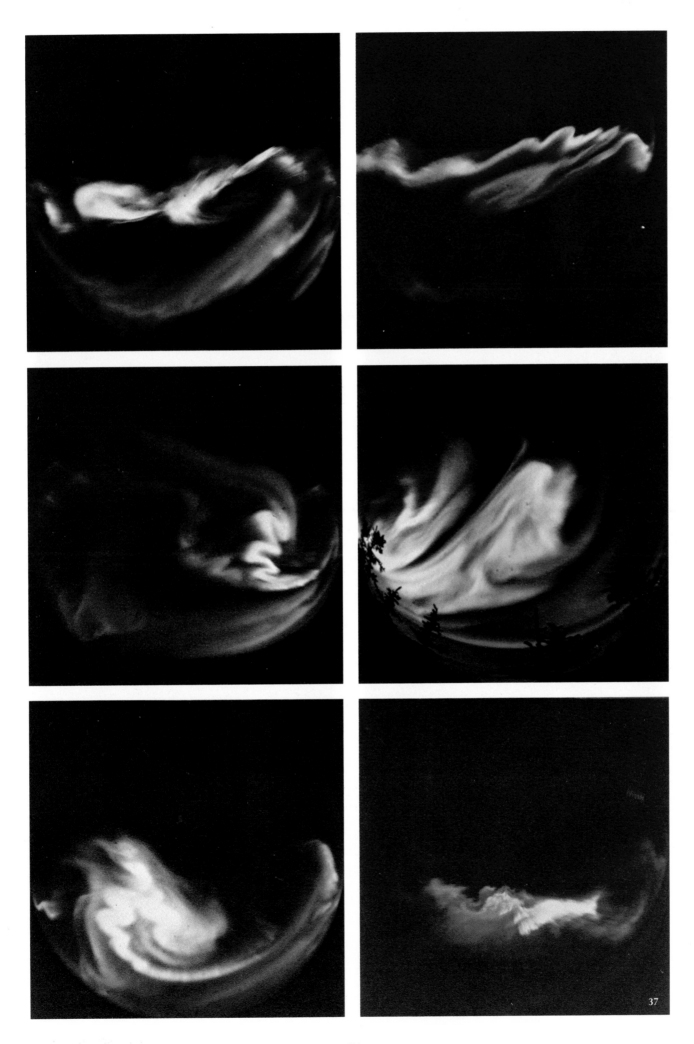

38

39

40

42

43

44

45

46

47

48

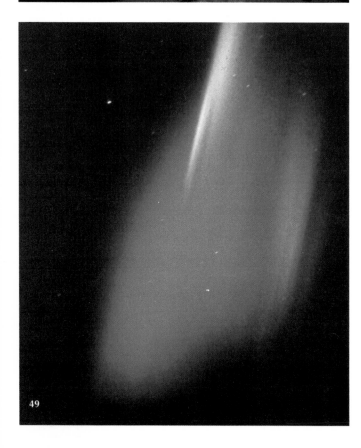

49

50

51

52

53

54

55

56

The timelessness of the aurora. This is a remarkable comparison of an aurora photographed near Fairbanks, Alaska, in the late 1960's and a painting by the Danish artist Harald Moltke from a Danish expedition to Iceland in 1899. The shapes, color, and structure in the pictures are almost identical.

QUIET

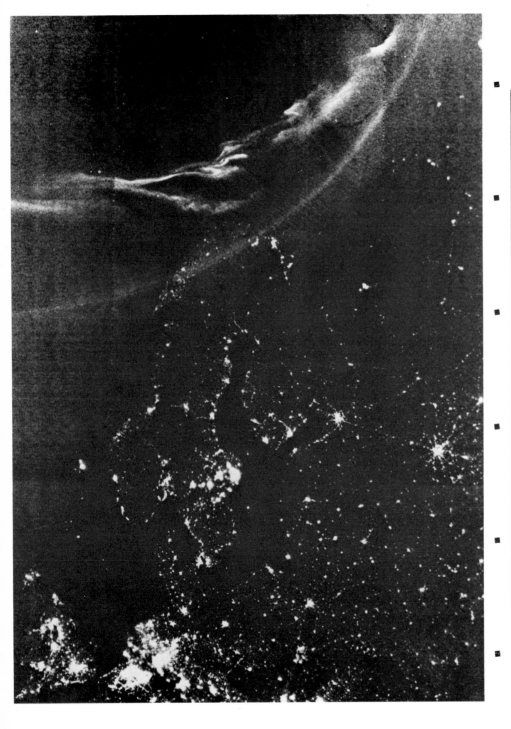

MODERA

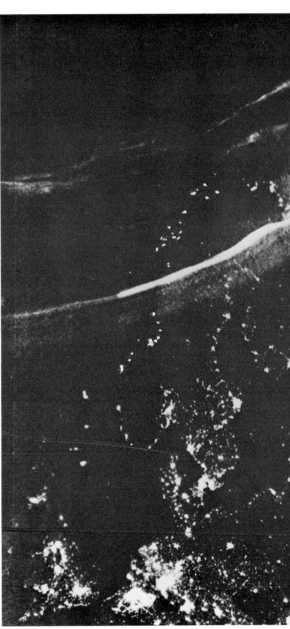

ACTIVE

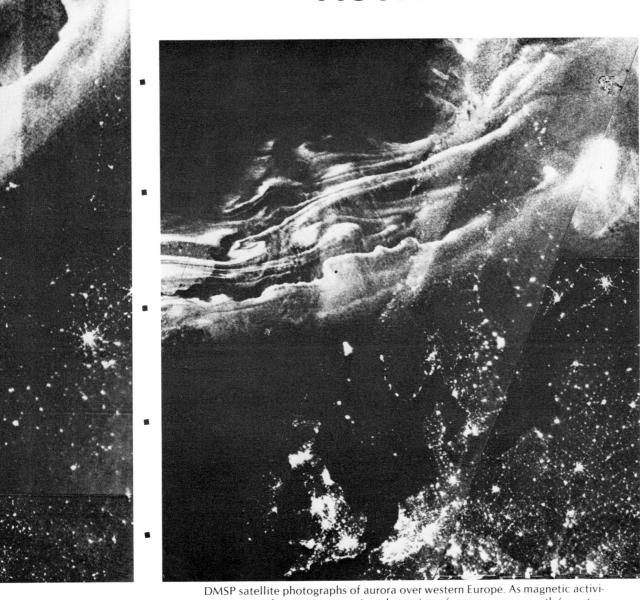

—70°N

—65°N

—60°N

—55°N

—50°N

DMSP satellite photographs of aurora over western Europe. As magnetic activity increases from quiet to active, the region of aurora moves south (equatorward), expands in latitudinal extent, and takes brighter, more complex forms (January 1979).

Following are the captions for Figures 1–56.

Figures 1–4, pages 8 and 9.

1 DMSP photograph of a quiet auroral arc and folded band over northern Scandinavia. City lights clearly outline western Europe (January 1979).

2 This is a composite of five pictures, each covering about a 30° field of view showing the full extent of an auroral arc over Mawson, Antarctica, in 1963. Arcs align along parallels of geomagnetic latitude.

3 Arcs of aurora australis photographed from Skylab when it was near 48.6°S, 68.8°E on September 11, 1973.

4 Quiet auroral arc near Fairbanks, Alaska.

Figures 5–12, page 10.

5 A rayed, curled band over Fairbanks, Alaska, in 1973.

6 A horseshoe-shaped rayed band over Mawson, Antarctica, in 1963.

7 Rayed bands near Fairbanks, Alaska, in 1973.

8 Quiet band through zenith near Fairbanks, Alaska, in 1973.

9 Part of fisheye view of active rayed band through zenith over Churchill, Canada, in 1972.

10 Part of fisheye view of folded band through zenith over Churchill, Canada, in 1972.

11 Rayed arc over Mawson, Antarctica, in 1963.

12 Individual auroral rays over Mawson, Antarctica, in 1963.

Figures 13–20, page 11.

13 Rayed band over Mawson, Antarctica, in 1963.

14 Curled band over Churchill, Canada, in 1972.

15 Folded band over Mawson, Antarctica, in 1963. In the foreground are two all-sky cameras.

16 Active band near Fairbanks, Alaska, in 1973.

17 Corona over Mawson, Antarctica, in 1963. The characteristic star pattern of the Southern Cross may be seen at the lower right.

18 A rayed band silhouettes the Chatanika radar facility near Fairbanks, Alaska, in 1973.

19 Two quiet arcs over Mawson, Antarctica, in 1963.

20 Active rayed arc over Mawson, Antarctica, in 1963.

Figures 21–26, pages 12 and 13.

21–26 A variety of auroral forms over Chatanika, Alaska, in 1973. These are individual frames from the film "Spirits of the Polar Night."

Figures 27–29, page 14.

27 Multiple rayed auroral bands near Fairbanks, Alaska.

28 Band near Fairbanks, Alaska.

29 Folded band through zenith over Ft. Churchill, Canada, in 1972.

Figures 30 and 31, page 15.

30 Beautiful fleeting colors appear in the corona over Fairbanks, Alaska.

31 A beautiful auroral corona over Kiruna, Sweden.

Figure 32, pages 16 and 17.

32 This is a time sequence (running column 1, top to bottom, column 2, top to bottom, etc.) from the film "Spirits of the Polar Night," showing the development of an auroral substorm. Total time covered by the 16 fisheye photographs (showing the whole sky from horizon to horizon) is 20 minutes. Churchill, Canada, March 17, 1972, near midnight.

Figures 33 and 34, pages 18 and 19.

33–34 Rayed bands near Fairbanks, Alaska.

Figures 35 and 36, page 20.

35 Six fisheye photographs, taken toward the south. Churchill, Canada, in 1972 at 8, 9, and 10 P.M. (top row) and 11 P.M., 12 midnight, and 1 A.M. (bottom row). The sequence shows typical behavior of southward-moving arcs (8–11 P.M.), a weak substorm (midnight), and diffuse postsubstorm veil (1 A.M.).

36 "Black aurora." Sometimes the complete sky is covered by a veil-type aurora except for thin regions of no luminosity. These regions move around and give the visual impression of a "black" aurora. Near Fairbanks, Alaska, in 1973.

Figure 37, page 21.

37 A selection of all-sky photographs of aurora from Ft. Churchill, Canada, in 1973.

Figures 38–41, pages 22 and 23.

38–41 A variety of bands and multiple band forms. All near Fairbanks, Alaska, except Figure 39, which is at Ft. Yukon, Alaska.

Figures 42 and 43, page 24.

42 A rare shot of a great red aurora associated with an extreme magnetic storm during the International Geophysical Year (Geophysical Institute, Fairbanks, Alaska, in 1958).

43 Subtle pink colors are seen in this rayed veil near Fairbanks, Alaska.

Figure 44, page 25.

44 Multiple bands near Fairbanks, Alaska.

Figures 45–50, page 26.

45 Folded band silhouettes totem pole, University of Alaska campus, Fairbanks, Alaska.

46 Curled bands near Fairbanks, Alaska.

47 Folded bands near Fairbanks, Alaska.

48 Rayed band near Fairbanks, Alaska.

49 Red rayed patch near Fairbanks, Alaska.

50 Rayed band near Fairbanks, Alaska.

Figures 51–56, page 27.

51 Rayed band at twilight over Kiruna, Sweden.

52 Rayed band near Fairbanks, Alaska.

53 Rayed band at twilight near Fairbanks, Alaska.

54 Pink rayed veil aurora near Fairbanks, Alaska.

55 Active band over Kiruna, Sweden.

56 Active band near Kiruna, Sweden.

2
The First 2000 Years

In the thirtieth year, in the fifth day
of the fourth month, as I was among the
exiles on the banks of the river Chebar,
heaven opened and I saw visions from God

Old Testament
Book of Ezekiel, 1:1, 593 B.C.

The aurora has been present in polar skies since before recorded history. By its location in these frozen unpopulated regions it has eluded the gaze of ancient and medieval writers and philosophers, so perhaps is the least written about and most neglected of all natural phenomena, yet it is without question the most fascinating and mysterious of nature's displays.

Occasionally, great aurora are seen at more temperate latitudes, but the rarity of such events is reflected in the paucity of historical references to the aurora. The men mentioned in the Preface may have witnessed an aurora only once or twice in their lifetimes. Auroras seen at mid-latitudes usually appear tired and sluggish in comparison to the colorful, crisp, and spectacular displays that transform the bleak polar skies.

It has recently been suggested (430) that our earliest record of auroras might be found in the serpentine meanders, or *macaronis,* of Stone Age man. These are found on rocks and cave walls from Cro-Magnon times and do not resemble anything normally associated with Stone Age existence. We know that there have been more than 30 reversals of the earth's magnetic field in the last 10 million years. Associated wandering of the magnetic poles would have resulted in periods when the aurora was seen much more frequently at mid-latitudes than is the case today. Consequently, it is not unreasonable to suppose that Stone Age man gazed with concern at flickering auroral lights and perhaps recorded them in his art. The most dramatic example of this possibility is to be found on the ceiling of the cave of Rouffignac in France, where lines drawn in the red clay ceiling resemble the folded curtain patterns of the aurora.

The Aurora in *Genesis?*

In *Genesis* 15 (the covenant with Abraham), God's promise is ratified in the ancient ritual of covenant. In this ritual the two parties walked between the halves of a dead animal and called down on themselves the fate of the divided animal should they violate the agreement. In *Genesis* 15, which describes the lands God gives to Abraham, Abraham kills and divides a heifer, a goat, and a ram; and when the sun descends and darkness has fallen,

> Behold a smoking furnace (oven)
> and a burning lamp (fiery torch)
> that passed between those pieces.
> *Genesis,* 15:17, ~2000 B.C.

The smoking furnace and burning lamp represent God walking between the divided animals to seal His promise in a way Abraham would readily understand.

If this oldest of stories of The Old Testament has any factual basis, it implies a heavenly illumination, described in some translations as a fiery torch passing across the night sky.

Might this be the oldest written account of an aurora?

The earliest references to the aurora according to some interpretations are four passages from *The Old Testament: Genesis* (15:17, ~2000 B.C.), *Jeremiah* (1:13, 626 B.C.), *Ezekiel* (1:1–28, 593 B.C.), and *Zechariah* (1:8, 518 B.C.). Of these, the first chapter of *Ezekiel* seems to be the clearest description of an auroral display:

> ...a stormy wind blew from the north, a great cloud with light around it, a fire from which flashes of lightning darted...(living creatures appeared)...between these animals something could be seen like flaming brands or torches, darting between the animals; the fire flashed light, and lightning streaked from the fire....Over the heads of the animals a sort of vault, gleaming like crystal, arched above their heads...(a human form appeared)...And close to and all around him from what seemed his loins upward was what looked like fire; and from what seemed his loins downward I saw what looked like fire, and a light all around like a bow in the clouds on rainy days...

To one familiar with auroral displays, this metaphorical description is nearly perfect, though Eric von Daniken in *Chariot of the Gods* (129) prefers an interpretation in terms of "a visit to earth by unknown intelligences of the cosmos." But if one examines the auroral photograph (page 37), it takes little imagination to see a man (or woman) kneeling and warming his hands over a fire, and it seems certain that Ezekiel witnessed a rare low-latitude aurora as he sat among the exiles on the banks of the Chebar on the fifth day of the fourth month of the year 593 B.C.

Perhaps the same aurora was seen in Greece by the philosopher Anaximenes, who lived at that time and wrote of "inflammable exhalations from the earth." Xenophanes also lived in the sixth century B.C. and mentioned "moving accumulations of burning clouds." These have been referred to as the first probable mentions of aurora in Greek literature (103, 448), though earlier Greek sky myths might also have been

Vidit Ieremias propheta in cœli nubibus ingentem uirgã, atqp ad cædēdum iam paratã: ollã pterea fuccē fam, & faciē eius (ut fcripturæ facræ uerbis utar) à facie aquilonis; et audiuit dñm dicentē: Vigilabo fuper uerbo meo, ut faciam illud, & ab aquilone pãdetur malũ fuper oēs habitatores terræ, quia ecce ego conuocabo oēs cogitationes regnorũ aquilonis, & uenient, ponentqp unufquifqp folium fuum in introitu portarum Hierufalē. Reliqua uide apud Ieremiam cap. 1.

And the word of the Lord came unto me the second time, saying, What seest thou? And I said, I see a seething pot; and the face thereof *is* toward the north.
Jeremiah, 1:13, ~ 626 B.C.
Though far from conclusive, some historians interpret this passage to record an auroral occurrence. This sixteenth century woodcut from Lycosthenes' *Prodigiorum* . . . shows Jeremiah's two visions of the rod of an almond tree, and the seething pot.

inspired by auroral displays. For example, Hesiod's *Theogony*, a creation story probably written in the eighth century B.C., records "flaming heavens" and "fiery sky dragons." The origin of the myth of Olympia might even be associated with the rare occurrence of auroras over the chain of mountains in northern Greece where the ancients located their Gods (306).

Julius Obsequens' book on prodigies (354) tells us "military spears were seen on fire in the sky late at night" in 502 B.C. This event is also recorded in a vast collection of ancient prodigies published by Lycosthenes in 1557 and repeated by Frobesius in 1739 in his catalog of *Modern and Ancient Spectacular Appearances of the Aurora Borealis*, where it is interpreted as the oldest aurora mentioned in classical literature. Subsequent compilers of catalogs of ancient auroras (Fritz in 1873; Schove in 1948; Link in 1962) also claim this to be the oldest auroral mention in classical literature, apparently accepting the earlier sources as reliable. Stothers (448) has recently shown that they are all incorrect; he traced the event to the *Roman Antiquities* of Dionysius (142), who gives further details that indicate a case of static electricity playing on iron spearheads.

In the fifth century B.C., Hippocrates and his student Aeschylus developed their theory of reflected sunlight (448), an erroneous explanation of the aurora which was to surface many more times over the ensuing 2,500 years. A description from 467 B.C. (probably from the lost works of Anaxagoras, preserved in part by Plutarch (372)) is the first Greek report that leaves absolutely no doubt that it concerns an auroral display:

> For seventy five days continually, there was seen in the heavens a fiery body of vast size, as if it had been a flaming cloud, not resting in one place, but moving along with intricate and regular motions, so that fiery fragments, broken from it by its plunging and erratic course, were carried in all directions and flashed fire, just as shooting stars do.
>
> Plutarch, *Lysander* 12:4

Aurora Over Byzantium

About 360 B.C., Philip of Macedonia laid seige on Byzantium (later Constantinople, presently Istanbul). It is told that Philip dug several tunnels under the walls of the city and that one night his troops stole within the fortified lines. But dogs barked and alerted the Byzantines, and Heaven especially favored them, as a sudden radiance in the form of a crescent streamed across the sky. The mysterious light showed them all things clearly, and the Macedonians were repelled. Perhaps this light was the aurora borealis. Certainly a typical auroral arc appears crescent shaped when viewed away from the zenith.

So striking was the incident that it was later commemorated on the coins of the city, and the crescent so represented may have suggested the famous symbol of Turkey. (524)

Flag of Turkey

A fish-eye photograph of the aurora, showing the whole upper hemisphere of the sky from horizon to horizon (Churchill, Manitoba, Canada, March 17, 1972, 12 minutes before midnight). One can readily see the form of a woman kneeling before a fire, lending support to the interpretaton of Ezekiel's vision as being an auroral display.

A fish-eye photograph of the aurora (Churchill, Manitoba, Canada, March 17, 1972, midnight). It is not difficult to see how primitive cultures could interpret such forms as a dragon in the sky.

Like a monster rous'd from sleeping,
First to westward slowly creeping
J. K. Foran
The Aurora Borealis

In the photograph, North is at the top and West is to the right.

37

Aristotle   384-322 B.C.
Greek philosopher

Ancient observations of the aurora in the Western world were made almost entirely from Greece and Rome, where modern data suggest the frequency of auroral occurrence would only be 1 to 3 times a decade. The auroral catalogs mentioned above identify eight or nine auroras prior to 300 B.C. A recent critical evaluation by Stothers (448) accepts a total of 55 probable auroral reports before the birth of Christ.

Aristotle in his classical treatise *Meteorologica* (29) was the first to provide a more dispassionate scientific description of auroras, and he gave them the name *chasmata*:

Sometimes on a clear night a number of appearances can be seen taking shape in the sky, such as chasms, trenches and blood-red colors.

Aristotle's science recognized four elements, fire, air, water, and earth; he theorized that heat from the sun caused a vapor to rise from the earth's surface and collide with the element fire which then burst into flames producing the aurora. He argued that reflection and attenuation of the fiery light as it passes through the denser, lower air contribute to the production of the observed colors, and he implied that the large angular extent and speed of the phenomenon confirmed its sublunary origin.

Aristotle clearly observed the aurora personally, so he probably witnessed the extensive displays of 349 and 344 B.C., though he may have seen other events during his years in Macedonia (343–336 B.C.), the latitude of which is more favorable for auroral occurrence than that of Greece.

From other regions around the Mediterranean, we have another biblical reference in the following passage from *Maccabees II* (5:2–3, 168 B.C.):

And then it happened that through all the city, for a space of almost forty days, there were seen horsemen running in the air, in cloth of gold, and armed with lances like a band of soldiers. And troops of horsemen in array, encountering and running one against the other, with shaking of shields, and multitude of pikes, and drawing of swords, and casting of darts, and glittering of golden ornaments and harness of all sorts.

The earliest Roman auroras were recorded in the histories of Livy and Dionysius and dated in the period 464–459 B.C. (448)—"the sky was seen to blaze with numerous fires." Pliny the Elder, the great Roman encyclopedist of the first century A.D., cataloged the "blood colored" aurora of Aristotle's time (349 B.C.) as a terrible portent. He described auroras (371) seen in Italy in 112 B.C. as follows:

Pliny (the Elder)   23-79 A.D.
Roman naturalist

We sometimes see, them which there is no presage of woe more calamitous to the human race, a flame in the sky, which seems to descend to the earth on showers of blood; as happened in the third year of the 107th Olympiad, when Philip was endeavoring to subjugate Greece.

Subsequently, auroras in Roman times were recorded rather matter-of-factly, except when the appearance of ray structure (6 times out of 40 accounts (180)) suggested a military association. Then links were drawn with major events in a superstitious way. Auroras are said to have appeared in the sky in 44 B.C. in the shape of armies of horse and foot before the death of Julius Caesar, which they were supposed to foretell

(62), and were seen over Palestine before the siege and destruction of Jerusalem by Titus Vespasian (251). However, Stothers (448) rejects this latter report as aurora and suggests the event was a static electricity phenomenon (St. Elmo's fire) that could have occurred during an intense lightning storm at night.

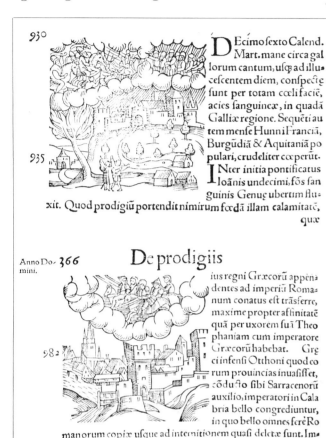

Auroras are said to have appeared in the sky in **44 B.C.** in the shape of armies on horse and foot foretelling the death of Julius Caesar. These later illustrations from Lycosthenes' *Prodigiorum . . .* depict auroras seen in 935 A.D. and 982 A.D. with similar imagination.

Seneca in *Naturales Quaestiones I* (414) expanded on Aristotle's work and wrote descriptions without superstitious flavor:

Several kinds are known: the *abysses*, when beneath a luminous crown the heavenly fire is wanting, forming as it were the circular entrance to a cavern; the tuns (*pittitae*), when a great rounded flame in the form of a barrel is seen to move from place to place, or to burn immovable; the gulfs (*chasmata*), when the heavens seem to open up and to vomit flames which before were hidden in its depths.

Auroras seen at Mediterranean latitudes are commonly red, so when seen low on the northern horizon, they were often mistaken for a major conflagration. Seneca tells us this happened in Rome in the time of Tiberius (37 A.D.) when

. . . the cohorts hurried to the succor of the colony of Ostia, believing it to be on fire. During the greater part of the night the heaven appeared to be illuminated by a faint light resembling a thick smoke.

This must have been a very important aurora, as the town of Ostia is to the southwest of Rome and auroras would only be seen there under extremely disturbed solar conditions.

Seneca 4? B.C.–65 A.D.
Roman dramatist, historian, statesman

Mitsuo Keimatsu 1907-1976
Professor of Oriental History, Kanazawa
University, Japan. He spent 13 years
searching old Oriental records for
mention of the aurora and catalogued
578 such references between 687 B.C.
and 1600 A.D.

D. Justin Schove, Principal of St.
David's College, Kent, England. He has
researched extensively historical
accounts of aurora in Europe and the
Orient.

Cicely M. Botley, Tunbridge Wells,
England. She has researched historical
accounts of the aurora.

Chu-Long, the candle dragon.

The fall of Rome marked the beginning of a period of a thousand years during which only a few references (usually fanciful or fearful) to the aurora are to be found in occidental literature, though numerous references are found in the carefully preserved records of the royal astronomical observatories in China and other oriental countries. Between the fifth and tenth centuries A.D. it is possible to find coincident sightings by month and even by day of auroras in Europe and the Orient (256). Reports in this period are usually poetic and exaggerated and followed by comments that such unusual phenomena in the heavens may predict new wars, agricultural disasters, or some other unhappy event.

It is difficult to identify the earliest mention of the aurora in ancient Chinese writings. A recent article on the subject (490) presents seven possible auroral references before 2000 B.C., the earliest in about 2600 B.C.:

> The mother of the Yellow Emperor Shuan-Yuan, Fu-Pao, saw a big lightning circulating around the Su star of Bei-Dou (Ursa Major α) with the light shining all over the field. She then became pregnant.
>
> *Chu-Shu-Chi-Nien*
> (Bamboo Album of Chronology)

An auroral interpretation of this message is based on the northern direction, a light able to illuminate a whole field, the fact that one would not expect to see stars during real lightning because of thunderstorm clouds, and the fact that lightning is a common metaphor for some auroral types. It was common in ancient times to associate rare sky happenings with the birth of important people.

In another ancient Chinese Book, *San-Hai Ching* (Book of Mountains and Seas), 2200 B.C., reference is made to a god in northern regions, described as a snake of red color who shines in the dark. His name is *Chu-Long* (candle dragon), and an auroral inspiration seems probable (490).

An extensive catalog of early aurora in China (256) begins with an event in 687 B.C., lists 18 additional reports up to 5 B.C., and lists no less than 170 events between the first and tenth centuries A.D. No particular name was given to the aurora through this period, and the phenomenon was described by familiar terms such as fire and animal names and most often was associated with the stars. One of these early Chinese descriptions (348) is from 30 B.C. when

> . . . at night there were seen in the sky luminous vapors, yellow and white, with streamers more than 100 feet long, which brightly lit up the ground. Some said that these were cracks in the heavens (thein lieh); others said that they were the swords of heaven (thien chiens).

This description is closely reminiscent of Aristotle's *chasmata* in his *Meteorologica,* but no connection between the two sources seems possible. Many of these early descriptions were collected in a 1652 volume that showed fanciful drawings of the aurora from China (459) (opposite).

In about the year 400 A.D., historians related (306) that columns were suspended in the sky over Europe for the space of 3 days and that "a fire burned behind a cloud which was terrible for its splendor, sometimes overspreading the sky."

Descriptions of auroras in the Middle Ages tended toward superstition and military significance. An example is the aurora seen in France on April 3, 451 A.D., and associated with the famous defeat of Attila during the summer of 451 A.D. at Catalaunic Field (150). Idatius Episcopus (394–468 A.D.) wrote (284)

> . . . after sunset, red streamers like fire or blood, intermixed with red shafts like spears, were seen from Acquilonis . . .

It is clear that the aurora shone forth on occasions during this period, the so-called Dark Ages of European history, but

> . . . at that time, the nations of Europe were sunk in ignorance and barbarism; and whatever phenomena the heavens presented were lost to posterity, from the rudeness and want of knowledge of the people of that age.

A century later, around 567 A.D., auroras seen in Italy were interpreted by the Benedictine Gregorius Magnus as omens announcing the invasion of the Lombards in 568 A.D. (284). In these accounts we find the first use of the words *acies igneae* ("battle lines of fire"), which were used either literally or with slight variations in descriptions of aurora in the following centuries.

Probably the earliest mention of aurora in England was by Mathew of Westminster (313) in 555 A.D. when he wrote of lances seen in the air, and one of the most spectacular auroras seen in Europe for over a hundred years seems to have been that of 585 A.D., recorded simply and accurately by Gregory of Tours (206):

> Two nights in succession we saw signs in the heavens, that is to say rays of light which rose in the north, as often happens . . . There was in the middle of the heaven a cloud of light, towards which all the rays converged, in the form of a tent of which the sides, much wider at the foot, ascended, narrowing to the summit, where they united, often in the shape of a hood.

The description of this and other auroras by Gregory of Tours is unusual for the period, as it considers the aurora as a curious manifestation and suggests no ideas of the supernatural. More commonly, astrology and events in the heavens so troubled the minds of men that the aurora borealis often became a source of terror, such as is indicated in the following report from England (247) taken from the *Anglo-Saxon Chronicles*, in 793 A.D.:

> This year dire forewarnings came over the land of the Northumbrians and miserably terrified the people; there were excessive whirlwinds and lightnings and fiery dragons were seen flying in the air.

There are many other references to auroras between 500 and 1100 A.D. in the *Chronicles of Scotland*, the *Anglo-Saxon Chronicles* (247), and the *Irish Chronicles*. The most common descriptions are of the sky burning and of fiery dragons.

The end of the first millenium A.D. saw the first authentic auroral observers—men who regularly studied the sky and reported on unusual happenings in a more factual way. Mathieu of Edesse recorded auroras in the Middle East in 1097, 1098, 1099, and 1100 A.D.; and Vysehrad reported auroras from Prague in 1128, 1132, 1138, and 1139 (284). Auroras were recorded in England from 1170 to the end of the twelfth century by

A Chinese manuscript from the year 1652 shows these representations of aurora seen in early China.

Gregory of Tours 538-593 A.D.
French historian

Gervasius and Neubrigensis (284). One of these auroras, which appeared soon after the death of Thomas à Becket in 1170, was seen by many as the blood of the martyr going up to heaven (62).

Since civilization began, auroras have always been most frequent at more northern latitudes and must have been a common sight to the Vikings in Norway. Consequently, one would expect a less superstitious attitude in Scandinavian countries. Indeed, more than a thousand years after the scientific descriptions by Aristotle and Seneca, an unknown Norwegian living to the north of Trondheim puzzled over a physical explanation for the aurora. He wrote a remarkable philosophical and political work in about 1250 A.D. called *Kongespeilet* ("King's Mirror") (273). The work is constructed as a dialogue between a wise and learned father and his son who is soon to go away on a long trip. The son asks many questions about the world he will encounter, and the father's replies give us one of our clearest insights into the state of culture and civilization in the thirteenth century. On the subject of the *nordurljos* the father says,

> . . . However, it is true of this subject [Northern Lights] as of many others of which we have no sure knowledge, that thoughtful men will form opinions and conjectures about it and will make such guesses as seen reasonable and likely to be true. But these northern lights have this peculiar nature, that the darker the night is, the brighter they seem, and they always appear at night but never by day, most frequently in the densest darkness and rarely by moonlight. In appearance they resemble a vast flame of fire viewed from a great distance. It also looks as if sharp points were shot from this flame up into the sky; they are of uneven height and in constant motion, now one, now another darting highest; and the light appears to blaze like a living flame . . .
>
> The men who have thought about and discussed these lights have guessed at three sources, one of which, it seems, ought to be the true one. Some hold that fire circles about the ocean and all the bodies of water that stream about on the outer sides of the globe; and since Greenland lies on the outermost edge of the earth to the north, they think it possible that these lights shine forth from the fires that encircle the outer ocean. Others have suggested that during the hours of night, when the sun's course is beneath the earth, an occasional gleam of its light may shoot up into the sky, for they insist that Greenland lies so far out on the earth's edge that the curved surface which shuts out the sunlight must be less prominent there. But there are still others who believe (and it seems to me not unlikely) that the frost and glaciers have become so powerful there that they are able to radiate forth these flames. I know nothing further that has been conjectured on this subject, only these three theories that I have presented; as to their correctness I do not decide, though the last looks quite plausible to me.

Thus in the father's opinion, cold is a positive force as much as heat or any other form of energy. To the men of the author's time, there was nothing strange in this belief; it seems to have been held by many even before the thirteenth century that ice under certain conditions could produce heat and even burn (273).

The last three centuries of the Middle Ages yielded little in the way of auroral reports. Besides being a period of less than average solar activity (and so few auroras would have appeared over Europe at any rate), it was an age when enquiry and reason were stifled by religious dogma. The return of auroras to European skies in the early sixteenth century resulted in the first printed description (224) of the aurora, by Creutzer, in the year 1527 (opposite). More frequent and spectacular appearances later in the century inspired a new analytical treatise (194) of the phenomena, published by Cornelius Gemma in 1575; it contained the first scientific illustration depicting the aurora.

The First Printed Auroral Description

The printing press was spreading through Europe in the latter half of the fifteenth century. By the end of the century, curious phenomena that occurred were sometimes described in pamphlets, often only a single sheet and sometimes as long as 10 or 15 pages. These pamphlets were usually illustrated with a woodcut.

The first such pamphlet (224) appeared in 1490, describing "lightning" seen from Constantinople on July 13. The description and accompanying woodcut might also be interpreted as reporting an aurora (224), but as auroral catalogs do not report widespread aurora on that date, such an inference seems unlikely.

About 20 other pamphlets appeared (describing six separate phenomena) before the first definite auroral description (224) of October 11, 1527. Many sources quote 1561 as the first printed auroral description, but even though the 1527 event was described by the general term "comet," there can be no doubt that it was an aurora. Three separate pamphlets were published, and the texts describe "a bent arm holding a great sword, blood red stars with streaks, mighty and frightening flames" in the sky. Auroral catalogs also report widespread aurora through Europe on this date.

The woodcuts from the three pamphlets are shown here. It is clear that one publisher copied from the other, making only slight changes in the illustration and text. Copies of the first two pamphlets may no longer exist (they were in the private library of G. Hellman in Germany in 1914, but two World Wars have occurred since then). The third pamphlet (in Latin) was available in the city libraries of Aachen and Wernigerode in 1914 and may still be preserved.

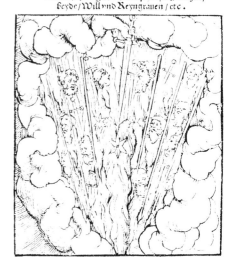

Title page from the 1527 pamphlet by Peter Creutzer. This was the first printed auroral description.

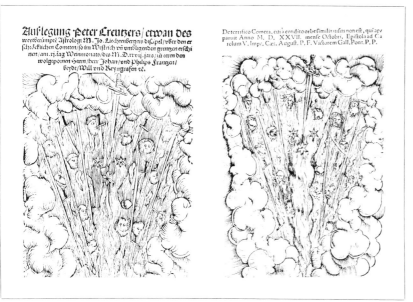

Two more pamphlets appeared soon afterward describing the same event. These were clearly copied from Creutzer with little change to the illustration or text. A fourth version appeared in Lycosthenes' *Prodigiorum. . .* in 1557.

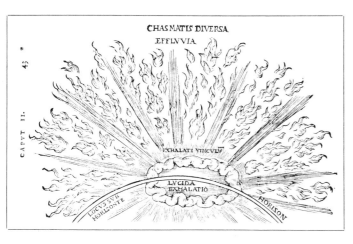

This diagram from Gemma's *De Naturae Divinis Characterismis* (1575) seems to be the first published scientific representation of the aurora.

William Fulke 1538–1589
English puritan Divinity fellow at Cambridge

In 1563, William Fulke, a puritan Doctor of Divinity at Cambridge, published a book (181) explaining almost all natural phenomena. It was entitled *A Goodly Gallerye With a Most Pleasant Prospect, Into the Garden of Natural Contemplation, to Behold the Naturall Causes of All Kynde of Meteors, as Wel Fyery and Ayery as Watry and Earthly* . . . In this remarkable book, Fulke presents natural scientific explanations without superstition. With regard to the aurora, he wrote (in part),

> According to their divers fashions, they have divers names: for they are called burning stubble, torches, dancing or leaping Goats, candles, burning beams, round pillars, spears, shields, Globes or bowls, firebrands, lamps, flying Dragons or firedrakes, painted pillars or broched steeples. . . The time when these impressions do most appear is the night season; for if they were caused in the day time, they could not be seen, no more than the stars be seen, because the light of the Sunne which is much greater, dimmeth the brightness of the being lesser. . . .
>
> Flying Dragons, or as Englishmen call them, fire Drakes, be caused on this manner. When a certain quantity of vapors are gathered together on a heap, being very more compact, as it were bardtepered together, this being of vapors ascending to a region of cold, is forcibly beaten back, which violence of moving, is sufficient to kindle it . . . then the highest part, which was climbing upward, being by reason more subtle and thin, appeareth as the Dragons neck, smoking, for it was lately in the repulse bowed or made crooked, to represent the Dragons belly. The last part by the same repulse turned upward, maketh the tayle, both appearing smaller, for it is further off, and also for that the cold hindeth it. This dragon thus being caused, flieth along in the ayre, and sometime turneth to and fro, if it meet with a cold cloud to beat it back, to the great terror of them that behold it . . .

But dispassionate writings such as these were still the exception rather than the rule, and the dragons and serpents in the sky that first appeared in English descriptions in the eighth century reappeared in a Czechoslovakian manuscript in 1571 (160):

> Fiery pillars were observed above the tower of Domzlice about the third hour of the night, and a dragon flying in the air above the whole town from the lower to the upper gate and even beyond the town. I have read somewhere and I have been told by my grandparents that this presages murders, fires, and other disastrous events.

Similar fears are expressed in the description (308) of the rare woodcut (opposite) of an aurora over Bohemia:

> In the year 1570, the 12th of January, for four hours in the night between midnight and sunrise, the portent appeared in the heavens after this fashion. At first a very black cloud went forth like a great mountain in which several stars showed themselves, and over the black cloud was a very bright streak of light, burning like sulphur and in the shape of a ship; standing up from this were many burning torches like tapers and among these stood two great pillars, one towards the east and the other due north, so that the town appeared illuminated as if it were ablaze, the fire running down the two pillars from the clouds above like drops of blood. And in order that this miraculous sign from God might be seen by the people, the nightwatchmen on the towers sounded the alarm bells; and when the people saw it they were horrified and said that no such gruesome spectacle had been seen or heard of within living memory . . . Wherefore, dear Christians, take such terrible portents to heart and diligently pray to God, that he will soften his punishments and bring us back into his favour, so that we may await with calm the future of our souls and salvation.   Amen.

The late sixteenth century was a period of more active solar activity, and many great auroras such as those of 1570 and 1571 appeared over Europe and were viewed by an uneasy populace. Just a few years later, Elizabethan England was startled by an aurora over London. In his *True and Royal History of Elizabeth, Queen of England* (88), Camden wrote

Aurora borealis near Bamberg, Germany, in December of 1560. Superstitious people explained the flashing lights in the northern sky as sparks from the clashing of swords in heavenly battles.

A shocking prodigy that has been seen from Kuttenberg in the kingdom of Bohemia and also in other places nearby on January 12, 1570, from 4 hours into the night lasting until 8.

Seen in Pressburg, Hungary, on February 10, 1681. It appeared in the early evening in the sky. It was witnessed by Paul Urbano, a respectable citizen of the town, and his whole household.

Sketches of aurora from the diary of Absalen Pederssön Beyer (1528–1575), teacher and minister in Bergen, Norway. These may be the oldest surviving sketches of the aurora.

. . . that in the month of November [1574], from the North to the South, burning clouds were gathered together in a round; the night following, the Sky seemed to burn, the flames running through all parts of the horizon, met together in the vertical point of Heaven. Nevertheless, let it not be imputed to me as a crime, to have made mention of these things in a few words, and by a short digression, since the greatest historians have recorded them in many words.

In September 1583 there arrived in Paris in formal procession, attired in the habits of penitents or pilgrims, 800 or 900 persons to present their gifts and ask prayers because of signs seen in the heavens and fire in the air (194).

Reports reemerging from Norway at this time are more descriptive and less fearful: Absalen Pederssën Beyer (1528–1575), teacher and minister in Bergen, kept diaries (preserved at the Royal Library in Copenhagen) in which there are a number of references to aurora. He describes flames, smoke, and narrow forms in his report of light in the sky on the night just before Christmas of 1563. Sketches in his diary may be the oldest surviving original representations of the aurora.

The end of the sixteenth century marks a natural end to the first chapter of the history of the aurora. This period saw a significant increase in auroral occurrence at European latitudes after some 300 years of rare sightings (presumably because of the minimal solar activity), and the time also coincides with the rapid increase in literacy in northern Europe. The Renaissance came to auroral latitudes later than to the Mediterranean, and with it came the beginning of real scientific enquiry into the physical nature of the aurora.

Aurora borealis seen from Nurnberg, Germany on October 5, 1591. The representation is in terms of heavenly fires.

3
The Beginning of Scientific Enquiry

And as imagination bodies forth
The form of things unknown, the
poet's pen
Turns them to shapes, and gives to
airy nothing
A local habitation and a name.

William Shakespeare
A Midsummer Night's Dream, 1600

From Man or Angel the great Architect
Did wisely to conceal, and not divulge
His secrets, to be scanned by them who ought
Rather admire. Or, if they list to try
Conjecture, he his fabric of the Heavens
Hath left to their disputes—perhaps to move
His laughter at their quaint opinions wide
Hereafter, when they come to model Heaven
And calculate the stars: how they will wield
The mighty frame: how build, unbuild, contrive
To save appearances; how gird the Sphere
With Centric and Eccentric scribbled o'er,
Cycle and Epicyle, Orb in Orb.

John Milton
Paradise Lost, 1667

Although sixteenth century Europe had seen the beginning of the great Renaissance of arts and sciences and also witnessed a marked increase in the frequency of auroras in European skies, this fortuitous combination of events did not immediately produce any scientific speculation as to the nature of the aurora. Fear and superstition continued into the early seventeenth century even among such learned men as physician Dr. Reyes de Castro Medice who described the aurora as a miracle (94) in a tract entitled *Prognostication of the Great Signs that Appeared in the Sky,* published in Spain in 1605. The following year though, a more enlightened Mercure Francois wrote (284), "In March and September some meteors and signs always appear in the sky and every physicist agrees that they bring neither good nor bad."

This period of history was nurturing some of the great astronomers; Johannes Kepler was just finishing his brilliant theory on the motion of the planets, and Galileo was assembling his first telescope. It was these men who provided us with more dispassionate descriptions of aurora: on November 17, 1607, Kepler minutely described a great aurora seen over Europe (306) and Galileo witnessed an outstanding display in Venice in 1621 (183). Galileo had already proposed an auroral theory reminiscent of Aristotle's; he described the aurora in a 1619 discourse (183) as

> . . . an effect which in my opinion has no other origin than that a part of the vapor-laden air surrounding the earth is for some reason unusually rarefied, and being extraordinarily sublimated has risen above the cone of the earth's shadow so that its upper parts are struck by the sun and made able to reflect its splendor to us, thus forming for us this northern dawn.

Gassendi, the French mathematician and astronomer, also described the aurora of 1621 in his textbook on physics (190). He reasoned that the vapor of the aurora must be at a great altitude, as the convexity of the earth was no hindrance to it being visible in the same situation from

Title page of a tract published in Spain in 1605, which interpreted the aurora as a miracle.

(Translation) Prognostication of the great signs that appeared in the sky on Thursday at 6 P.M., the day of the glorious martyrs Cisilo and Victoria, Patrons of Cordoba, on the 17th of November of the year 1605. Reyes de Castro Medice.

PRONOSTICACION.

DE LAS GRANDES

SEÑALES QVE PARECIERON en el Cielo Iueues a las seys de la noche, dia de los gloriosos Martyres Cisclo y Victoria, Patrones de Cordoua, a diez y siete de Nouiembre de mil y seyscientos y cinco Años.

Compuesto por el Doctor Reyes de Castro Medico.

SOL

Impresso con licencia en Cordoua. Año de M. DC. V.

48

places remotely distant from one another. He has been credited widely with introducing the term "aurora borealis" or northern dawn, but it appears that the credit must be given to Galileo (431) (see page 51). Auroras seen in the southern sky were for a time referred to as "aurora australis," obvious confusion being the result. The two terms later came to be applied to auroras seen in the northern and the southern hemisphere.

The great French philosopher René Descartes may also have seen the aurora around this time or at least had the phenomenon described to him. He seems to have been the originator of the idea that the aurora resulted from the scattering of sunlight from ice particles (139). These particles would be characteristic of the cold high-latitude regions where aurora were more common, and as they were high in the atmosphere, they could reflect sunlight even after the sun had passed below the horizon. This erroneous explanation was to resurface many times over the next three centuries.

Writings by scientists, however, had to compete with the more popular journals of the day, which continued to promote superstitious interpretation. In particular, the journal *Theatrum European,* which first appeared in 1618 and yearly thereafter, related all noteworthy events, politics, catastrophes, murders . . . and auroras too. The more commonplace tastes of the majority of readers were catered to by numerous auroral descriptions in medieval style.

Pierre Gassendi 1592-1655
French philosopher

René Descartes 1596-1650
French philosopher

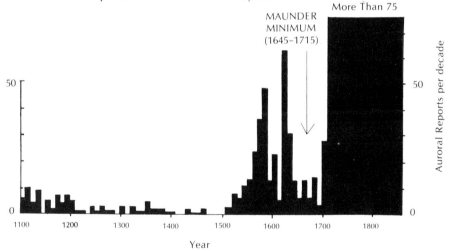

Year

This graph shows the number of reports of aurora at latitudes below the arctic circle (66°N) from 1100 to 1850 A.D. The rapid rise of auroral reports after 1500 coincides with improvements in communications (the printing press), increases in population, and an increase in the awareness of science. The marked minimum in this increasing trend between 1645 and 1715 is a real effect and corresponded to an almost complete absence of solar activity (as evidenced by sunspots). This period, known as the Maunder minimum, was named after the superintendent of the Solar Department of Greenwich Observatory, who was one of the first to stress the reality of this period of prolonged solar quiet.

After the 1620's, the remainder of the seventeenth century was characterized by almost a complete absence of solar activity (the so-called Maunder minimum (315)), and, as a result, there were few auroras to be seen at central European latitudes. Even as far north as Norway, displays were infrequent, and an unusual aurora of 1657 was interpreted as an omen (376):

A few years since died here Kemberg, in his 92nd year, our learned and experienced physician Abrose Rhoden who, while professor of natural philosophy and mathematics, at Christiania in Norway, predicted from the appearances which were observed at Eger in Norway on August 1, 1657 that Frederick III, who was then on the throne of Denmark, would be invested with an unlimited sovereignty, and that the kingdom before elective, would be thus made hereditary.

E. W. Maunder 1851-1928
English physicist

(And indeed, within 3 years after the prediction, the government was changed into an independent hereditary monarchy.)

This long period of low solar activity was briefly broken in 1660-1661, and auroras over London in March 1661 were viewed with some disquiet (405). A pamphlet was published that described the event, with a magnificent title page:

> Strange News from the West, being a true and perfect account of several Miraculous Sights seen in the Air Westward, on Thursday last, being the 21 day of this present March, by diverse persons of credit standing on London Bridge between 7 and 8 of the clock at night. Two great Armies marching forth of two clouds, and encountering each other; but, after a sharp dispute, they suddenly vanished. Also, some remarkable sights that were seen to issue forth of a cloud that seemed like a mountain, in the shape of a Bull, a Bear, a Lyon, and an Elephant with a Castle on his back, and the manner how they all vanished.
>
> London, Printed for J. Jones, 1661

This seemed the most spectacular of a handful of auroras that occurred around 1661 (a small sunspot cycle peak in the overall 100 year Maunder minimum period). It was perhaps these events that revived auroral studies briefly. The center of activity seemed to be Leipzig, Germany, where at least two doctoral students wrote their dissertations on this subject: Christopher Henry Starck wrote a thesis entitled "De Aurore Boréale" in 1663 and Nikolaus Daneil Frueauff submitted "De Aurore" in 1675 (223).

Edmund Halley lived during this period of the Maunder minimum. Although he was more famous for his comet studies, he was also fascinated by the historical accounts of the aurora and was, as he says, "dying to see the aurora and expected to die without seeing it." This did not deter him from theorizing about its probable cause: at first he suggested "watery vapors, which are rarefied and sublimed by subterraneous fire, might carry along with them sulphureous vapors sufficient to produce this luminous appearance in the atmosphere." Another hypothesis of Halley's was even more fanciful:

> Supposing the earth to be concave, with a lesser globe included, then, in order to make that inner globe habitable, there might not improbably be contained some luminous medium between the balls, so as to make a perpetual day below. And if such a medium is enclosed within, what should hinder us from supposing that some of this lucid substance may, on rare and extraordinary occasions, transude through and penetrate the shell of our globe, and being loose, present us with the phenomena above described. This seems favored by one circumstance, the figure of the earth; for that being a spheroid flattened at the poles, the shell must be thinner at the poles than in any other part, and therefore more likely to give passage to these vapors; whence a reason is derived for the light being always seen in the north.

At 60, Halley finally had the opportunity to witness personally an aurora (on March 17, 1716) when the finest aurora in a century appeared over Europe. At the request of the Royal Society he published "some conceptions ... proposed ... as seeming to some to render a tolerable solution of the very strange and surprizing phenomena thereof" (210).

He abandoned the above two hypotheses, which he evidently realized were insufficient to account for the phenomenon, and turned to effluvia of a more subtle nature such as "magnetical effluvia, whose atoms freely permeate the pores of the most solid bodies, meeting with no obstacle from the interposition of glass or marble or even gold itself." Halley pic-

The Origin of the Name *Aurora Borealis*

Galileo Galilei 1564-1642
Italian astronomer and physicist

Almost every article published in the last 250 years dealing with the history of the aurora attributes of the origin of the metaphor *aurora borealis* (northern dawn) to Gassendi, the French mathematician and astronomer, in the year 1621. Although Gassendi described an unusual aurora he witnessed that year, his first published use of the term *aurora borealis* was not until 1649. George Siscoe at the University of California at Los Angeles has recently researched the origin of the term (431) and concludes that it dates to 1619 in a work associated with and probably written by Galileo. Siscoe's research is summarized below.

The period between the 1620's and 1716 A.D. was one of very low solar activity (the so-called Maunder minimum), and very few auroras were seen in Europe during those years. A major aurora ended this state of affairs on March 17, 1716, and was described by Edmund Halley (210) in *Philosophical Transactions* that same year. In his article, Halley referred to the aurora of 1621 as "described by Gassendus in his *Physics*, who gives it the name *Aurora Borealis*." This mistaken credit seems to be the origin of error by later authors who wrote of auroral history.

A reading of the first auroral textbook (306) by Mairan (*Traité Physique et Historique de l'Aurore Boréal,* 1733) gives evidence of a prior origin, where on page 95 he wrote

> . . . this name [*aurora borealis*] that is generally believed to be imposed by the famous Gassendi, I will easily prove by Gassendi himself that it must have been given before him.

Mairan refers to Gassendi's statement in his *Syntagmatis Philosophici:* in the chapter on aurora he speaks of that light in the north which imitates dawn and which "some people name *aurora borealis*" and definitely implies an earlier (unspecified) source.

Mairan, in the second edition of his *Traité Physique et Historique de l'Aurore Boréale* (1754) suggests that earlier sources may go back as far as Gregory of Tours (538-594 A.D.) (206). Gregory wrote, "the heavens in the northern quarter (*septentrionali plaga*) shone brightly in this manner, as producing an imaginary dawn (*auroram*)." Gregory used the proper Latin word for northern, *septentrionali*, rather than the latinized Greek *borealis*, but the image is fully drawn a millenium prior to 1621.

But Gassendi does not reference the writings of Gregory in any of his works, so the northern dawn metaphor may have independently reappeared in the early seventeenth century. The earliest seventeenth century reference seems to be in articles concerning comets, published by Galileo and his student Guiducci. Galileo regarded auroras and comets as having similar origins, and in 1616

> . . . the sky at nighttime illuminated in its northern (*settentrione*) parts in such a way that its brightness yields nothing to the brightest dawn (*aurora*) and closely rivals the sun . . . thus forming for us this northern dawn (*questa boreale aurora*).

Siscoe points out that in this literal translation the expression is presented as a metaphor rather than as a name, but in a later work by Galileo (*The Assayer, 1622*), it is used as a name, and *aurora borealis* and *aurora* occur in at least eight places.

Letters written by Gassendi in 1640 show he was aware of Galileo's papers on comets and auroras, so this is probably where the name *aurora borealis* was first suggested to him, though he did not use it in print until 1649. It thus appears that either Galileo or his disciple Guiducci should be given credit for the introduction of the name *aurora borealis* into the recent scientific literature.

tured a circulating pattern of these magnetical effluvia along the lines of magnetic forces on the earth and illustrated it with a diagram of a magnetized sphere.

> ... subtle Matter freely pervading the Pores of the Earth, and entering into it near its Southern Pole, may pass out again into the Ether, at the same distance from the Northern . . . otherways discovering itself but by its effects on the Magnetick Needle, wholly imperceptible and at other times invisible, may now and then, by the Concourse of several Causes very rarely coincident, and to us as yet unknown, be capable of producing a small degree of Light . . . [as] . . . themselves become luminous, or rather may sometimes carry with them out of the Bowels of the Earth a sort of atom proper to produce Light in the Ether.

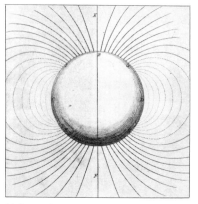

Halley's diagram of the dipole magnetic field of the earth used to illustrate his theory of the aurora.

Edmund Halley   1656–1742
English astronomer

Halley discussed the corona shape of the aurora, where rays of light appear like the spokes of a wheel converging towards a central point near the zenith. He made the remarkably astute observation that in all likelihood it was an effect of perspective, which is indeed the case. (Narrow auroral structures (rays) align along the magnetic field direction so that when viewed from directly underneath they appear to converge just as parallel railway tracks appear to converge in the distance.) Three years later, in 1719, Halley observed another aurora (211) and measured the elevation of the coronal point as 14°S of zenith, but curiously did not draw attention to the agreement of this direction with the local magnetic field in London. (It was left to the Swedish scientist Wilcke to draw this connection some 50 years later (501).)

Halley appears to have been the first to suggest a practical way to measure the height of aurora. He pointed out that if the auroral position had been noted accurately among the stars at two places, such as London and Oxford, whose distance apart was well known, by its *diversitas Aspectiss* the height might be determined. (Such measurements were more or less successful later in the century.)

These scientific discussions failed to convince believers in prodigies, as was evidenced from the title of a tract (336) written soon afterward entitled *An Essay concerning the late Apparition in the Heavens on the sixth day of March. Proving by Mathematical, Logical and Moral Arguments, That it cou'd not have been produced meerly by the ordinary Course of Nature, but must of necessity be a Prodigy. Humbly offer'd to the consideration of the Royal Society, London: 1716.*

This spectacular aurora seen over Europe was not seen all around the world. In America, New Englanders had to wait until 1719 to be startled by a great aurora in their skies (see page 94). This aurora was also seen in China (where one had already been reported in 1718) and caused so much amazement that engravings of them were struck off in the thousands and were secretly (portents being contraband inside the Great Wall) distributed through the empire (150).

The great aurora of 1716 marked the end of the Maunder minimum period. Auroras were seen quite regularly and were reported in contemporary scientific journals such as the *Philosophical Transactions of the Royal Society*. The first serious disagreement with Halley's views was published by Mr. W. Derham (F.R.S.) in 1728 (136). After a detailed description of auroras he had observed, he advanced his conviction that they were of the same matter or vapors that produced earthquakes. As evidence of the association, he quoted an auroral report of November 14, 1574, which was followed by a great earthquake on February 26,

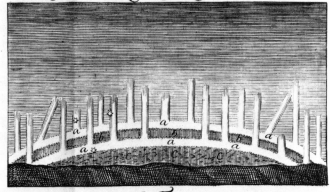

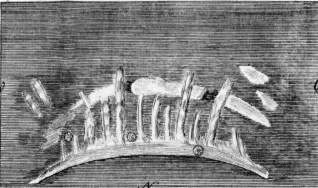

2. Figur:
Nordlÿset, seet i Kiöbenhafn ⅆ.1.og 6.Martii 1707.

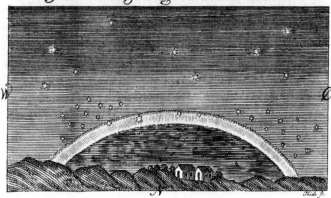

4. Figur: Nordlÿset, seet i Giessen udi Land-Greve
skabet Hessen ⅆ.26.Novemb.Aᵒ 1710.

Plates from an article on historical and physical accounts of the aurora, written by 'J.F.R.' and published in *Acta Societatis Hafniensis* in Copenhagen in 1745. It may be seen that the auroras of 1707 and 1710 were observed as quiet arcs, sometimes with ray structure, whereas the great aurora of 1716 exhibited much more spectacular forms. (See page 54.)

1575, and noted that the aurora he observed on October 8, 1728, was preceded by a fatal earthquake in Sicily and succeeded by one in England on October 25. He concluded that the northern lights are "of great use to the peace and safety of the earth, by venting some of that pernicious vapor and ferment that is the cause of these terrible convulsions, which earthquakes are accompanied with."

In France, Maraldi expressed the opinion (309) that the aurora was caused by "subtle, sulphurous exhalations," which contributed to calming the air, as auroras seemed to occur when the air was calm and temperate.

Aurora and Politics

The great aurora of March 17, 1716, marked the return of aurora to European skies after an absence of some 100 years. In England it was a time of civil disturbances, and the event inevitably took a party tinge, with Whigs and Tories wrangling over the meaning of these streaming lights in the heavens (389). The Jacobites muttered that such portents boded ill for the new dynasty; they talked of giants with flaming swords, fiery dragons and embattled armies. The more imaginative swore they heard the reports of firearms and smelled the powder burnt by the aerial combatants. The Hanoverians' line was to make light of the whole affair as a mere natural phenomenon. The journal *Flying Post* remarked that "the disaffected party have worked this up to a prodigy, and interpret it to favour their cause"; the writer then proceeded to dismiss it as sulphurous exhalations, kindled vapors, and will-o'-the-wisp coruscations from the fens (150).

10. Figur:
Nordlyset seet i Giessen d: 17 Februarii A: 1721.
Num: 1.

11. Figur:
Nordlyset seet i Giessen d: 1 Martii A: 1721.

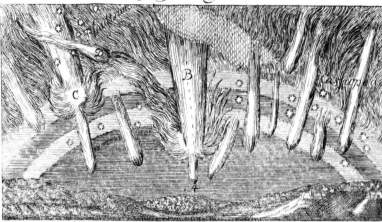

Num: 2.

12. Figur. Nordlyset, seet til Brevilleponti Frankerige d: 26. Septembris A: 1726

Thiele sc.

9. Figur
Nordlyset seet i Dantzig d: 17. Martii Anno 1716.

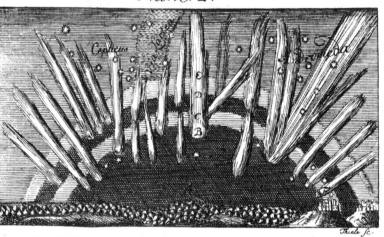

Num 1.   Num 2.   Num 3.

Num 4.   Num 5.   Num 6.

Thiele sc.

in auroral physics
ly to the subject,
ean Jacques d'Or-
published by the
on was printed in
the aurora was a
d also criticized
which related the
n extension of the
sed the outer por-
ons about the sun
uced the various
ar periodicity of
possible connec-
eturn of sunspots

ied in Paris in the
gh the author who
ensson) was a far
u Boréal, devotes
fects of the sun in
it was claimed that
e autumn and winter
ould catch fire and create the aurora.

Jean Jacques Dortoùs de Mairan
1678–1771
French natural philosopher

Atmospheric air movements were also thought to be responsible for aurora by Captain Heitman, a Norwegian naturalist. In 1741 his son published (222) a posthumous piece of his father's entitled *Physical Considerations About the Hot Sun, the Cold Air, and the Aurora,* which advanced the following theory:

> In the frigid zone . . . sometimes the lower regions of the air, which is filled with nitrous vapours, is whirled around, and then is formed that light in the air called the Aurora Borealis: yet this is a light devoid of heat . . . frequently happens at an approaching alteration of the weather . . . but only at particular seasons, when the saline corpuscles of the air are agitated by a natural fermentation . . . and when, by the elevation of the inferior air . . . compressed against . . . the lower part of the cold region, this causes those corruscations in the air, which are called the north light.

Leonhard Euler 1707–1783
Swiss mathematician

Mairan's book was the most widely circulated work on the aurora at this time, but his radical new theory was not widely accepted. Mairan was upset that Halley, whom he greatly respected, never gave his approval (though Halley never argued the matter publicly). A serious challenge to Mairan came in 1746, when the celebrated Leonhard Euler wrote a treatise to refute Mairan and establish a new doctrine of his own (165). He ascribed the aurora to particles from the earth's own atmosphere driven beyond its limits by the impulse of the sun's light and ascending to a height of several thousand miles. Near the poles, these particles would not be dispersed by the earth's rotation. He explained the zodiacal light and the tails of comets by the same hypothesis. Mairan was less than overjoyed by Euler's article, and in 1747 he complained (264), "Notice, if you please, that this is not the only occasion upon which M. E. has expressly attacked me—I who never pronounced or wrote his name except to sing his praises. I resent both the honour and consequences of them, and I really wish that he would just leave me in

TRAITÉ

PHYSIQUE ET HISTORIQUE

DE

L'AURORE BORÉALE.

Par M.ʳ DE MAIRAN.

Suite des Mémoires de l'Académie Royale des Sciences.
ANNÉE M. DCCXXXI.

A PARIS,

DE L'IMPRIMERIE ROYALE.

M. DCCXXXIII.

Title page from the first auroral textbook by Jean Jacques
Dortoùs de Mairan, published in Paris in 1733.

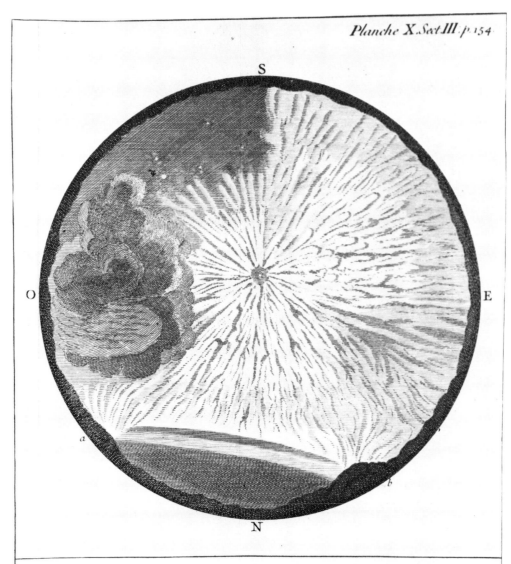

Fig. XVII. Aurore Boreale du 19.me Octobre 1726. telle qu'elle parut dans tout l'Hmisphere
Superieur du Ciel, vers les 8. heures du Soir, à Breuillepont, Diocese d'Evreux, 15 ou 16 Lieues
à l'Occident de Paris.

A figure from Mairan's book, illustrating a coronal form. The picture shows the full upper hemisphere of the sky, a forerunner of modern all-sky camera photographs.

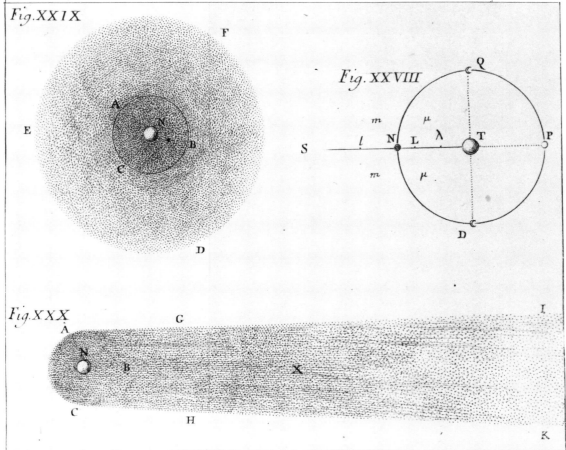

Plates 28-30 from Mairan's book, illustrating his theory relating the aurora to an extension of the sun's atmosphere, which mingled with the earth's atmosphere to produce the various phenomena of the aurora.

Simonneau Sculp.

Title page and plate from a book by
E. Sguario describing the aurora seen
from Venice on December 16, 1737.

DISSERTAZIONE
SOPRA
LE AURORE BOREALI,
DOVE

Con fiftema particolare fondato fopra i NEWTONIANI principj,
fopra le Leggi della MECCANICA, e fopra le migliori, e
più accurate offervazioni fi tratta delle medefime, e dove fi
riferifce principalmente la Storia, e le cagioni dell' Aurora, veduta qui in VENEZIA li 16. Dicembre verfo le ore 2. della
notte nell' Anno MDCCXXXVII.

DI

EUSEBIO SGUARIO
VINIZIANO.

DOTTORE IN FILOSOFIA, E MEDICINA.

IN VENEZIA, MDCCXXXVIII.
Appreffo PIETRO BASSAGLIA,

All' Infegna della Salamandra.
CON LICENZA DE' SUPERIORI, E PRIVILEGIO.

Aurora Boreale veduta da Venezia il di 16 Decembre dell'anno 1737.
siccome comparve alle 3 ore della notte.

peace." He vigorously defended his theory in a paper published that same year and continued to criticize Euler and promote his ideas, culminating in a second revised edition of his book in 1754. Euler never answered Mairan, content that his explanation was correct.

Thus the aurora featured in one of the great early debates among scientists during the Enlightenment period. J. Morton Briggs concludes (78),

> If Halley, Mairan, and Euler had flaws, it was not so much in their science as it was in their egos. Mairan was much too verbose and extraordinarily tiresome in his claim to have determined the cause of the aurora. While Euler and Halley were more sensible, it seems clear that each thought he was right.

Another who was not entirely impressed by Mairan's book or by Euler's ideas was the Right Reverend Erich Pontoppidan, Bishop of Bergen, in Norway. National pride might partly have motivated the following comment in the preface of his important book, *The Natural History of Norway* (376), written in 1751:

> Had M. de Mairan taken care to procure from Norway some accurate observations of the Aurora Borealis, his valuable *Traité Physique et Historique de l'Aurore Boréale* would have been much more complete and decisive; for the north light takes its rise from Norway, and particularly from the diocese of Drontheim.

Reverend Erich Pontoppidan 1698–1764
Danish historian and onetime Bishop of Bergen, Norway

Pontoppidan discussed the ideas of the celebrated Wolfius (514), who thought that the aurora "is a substance as yet immature for lightning . . . an imperfect tempest," which Pontoppidan comments may be corroborated by the crackling sound sometimes reported to accompany the aurora. The learned bishop had no time for theories attributing the north light to reflections of sunlight from high vapors when the sun is below the horizon, and this caused him to ridicule audaciously the ideas of Euler. He stated that Euler's explanation required that

> . . . first, there must be vapours in the upper region of the air; next, some clouds of that sort, and these at a vast height, and in the north; and they must not only emit vapours, but be illuminated and radiated from the sun, when it is invisible to us . . . and lastly, there must be a north-wind in the same upper region of the air to set it in motion, and to give a disposition to the figures, which so suddenly change their appearance.

The Bishop ironically concluded his discussion of Euler's ideas with, "It is possible that the experience of posterity may suggest something more probable."

Though disclaiming that physics was ever his chief study and that he "be far from the presumptuous conceit of believing to have discovered the secret designs of the infinite Creator," Pontoppidan neverthless proceeds to add his own opinion to the problematical subject. Electrical experiments had been known for 20 years, and Pontoppidan describes in detail an experiment (138) by M. Desaguliers reported in *Philosophical Transactions* in England, purporting to show that air itself is electrical. The experiment actually generated static electricity upon a rotating glass globe, which Pontoppidan likened to the rotating earth, suggesting that the north light

> . . . is the very ether itself; which, being aggregated, gives way to the impression of the humid air, and mounts and floats above the clouds, whose motion likewise renders it variable.

Anders Celsius 1701-1744
Swedish astronomer

Henry Cavendish 1731-1810
English scientist

Thus he says that aurora will not occur in the dry air of winter's frost or summer's heat but only when the weather is damp does "the north light break forth, as a certain prognostic of the change."

Perhaps the most accurate observation the Bishop made was that the general region of aurora in Norway was not due north but rather northwest. Unaware that this is an alignment around the magnetic pole rather than around the geographic pole, he uses the observation to support his theory, claiming that the rotating earth would cause the air to be rarefied on one side and compressed on the other; this would also explain lesser aurora in the morning hours.

Soon after the invention of the discharge tube in 1753, it was speculated that the aurora was a discharge phenomenon in the upper atmosphere (89). An almost universal conviction developed among scientists that the newly discovered electricity was the true cause of aurora, and for a long time there was little further speculation on the subject. Experiments with discharge tubes became a central part of auroral research for the next hundred years and are described in Chapter 16, *Man-Made Aurora*.

The eighteenth century saw considerable refinement of observational techniques. The connection between aurora and magnetic disturbances was discovered in 1741 and is usually attributed to Celsius (96), from Uppsala, Sweden. But one wonders if Celsius was quite fair to his student Hiorter, who wrote (233),

> Who could have thought that the northern lights would have a connection and a sympathy with the magnet . . . The first time I saw an aurora to the south and noted simultaneously a great movement of the magnetic needle was on March 1, 1741, in the evening . . . When I announced this to the professor [Celsius] he said that he had noted such a disturbance of the needle in similar circumstances, but had not wished to mention it in order to see . . . whether I, too, would light on the same speculation.

In 1790 the great English scientist Cavendish (95) used triangulation to estimate the height of auroras as between 52 and 71 miles. Halley and Mairan had described this triangulation technique earlier: two observers separated by a distance of tens of miles measure the elevation of the aurora, and knowledge of the distance between observers and some simple trigonometry then yields the auroral height. Unfortunately, it was not always possible to ensure that both observers were looking at the same part of an aurora, and as observations proliferated, the auroral height became more uncertain, reports varying from 5 to 1000 miles. The situation was not resolved finally until the beginning of the twentieth century, when the invention of the telephone allowed communication between observers and the new technology of photography permitted detailed analysis in the later comfort of a warm laboratory. (Average auroral heights are 100–120 kilometers.)

John Dalton, the famous English chemist, was also an avid observer of all meteorological phenomenon and became enamored with the aurora borealis. In 1792 he wrote this description (128) of a beautiful display seen from Kendal, near Manchester:

> . . . the whole hemisphere was covered with them, and exhibited such an appearance as surpasses all description. . . . The intensity of the light, the prodigious number and volatility of the beams, the grand intermixture of all the prismatic colours in their utmost splendor, variegating the glowing canopy with the most luxuriant and enchanting scenery, affording an awful, but at the same time, the most pleasing and sublime spectacle in nature.

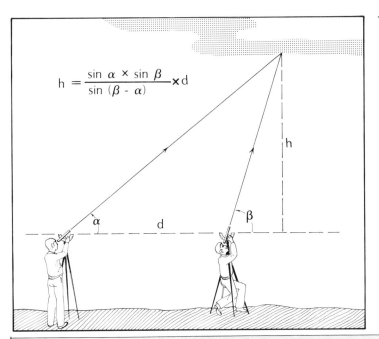

$$h = \frac{\sin \alpha \times \sin \beta}{\sin (\beta - \alpha)} \times d$$

The height of aurora may be determined by measuring its elevation from two points a known distance apart. This technique is called triangulation.

Meteors

When reading historical works on the northern lights, confusion can result if one is not aware of the old use of the word *meteor*. Often the term is used when it is clear that the aurora is being described. This is because it has only been during the past 100-150 years that *meteor* has taken on its modern meaning of a "shooting star." Before that, the word was used in a very general sense to include almost any phenomenon in the sky. The definition given for students of science by Peter van Musschenbroek (342) in his *Elements of Natural Philosophy* (published in 1744) was that,

> All bodies that are above us between the heaven and the earth, which are suspended, float, move, are driven, set on fire, conjoined, separated, ascend, descend, and produce any phenomenon whatever, we call *meteors*.

This plethora of phenomena included under the general term meteor must have often led to confusion. It went so far that in one case (*Philosophical Transactions of the Royal Society*, vol. *35*, p. 204, 1729) an article titled "On the Meteor Called Ignis Fatuus" referred to the firefly.

Peter Musschenbroek 1692-1761
Dutch mathematician and astronomer

Dalton became a conscientious auroral observer and made careful measurements with a theodolite on the position of auroral forms. He engaged the cooperation of a friend to try to measure the auroral height by triangulation, deciding on 100–150 miles, which convinced him that the phenomena lay above the earth's atmosphere, which he believed was confined to 15–20 miles. However, he cautioned as to the complete accuracy of triangulation height measurements because of "the great difficulty to ascertain that the observations were contemporary, and made upon one and the same object." But his theodolite measurements did prove that the aurora aligned parallel to the magnetic meridian and that luminous beams were "parallel to the dipping needle"; this convinced him of the magnetic nature of the phenomena (128). He thus concluded,

> . . . the beams of the aurora borealis are ferruginous in nature, because nothing else is known to be magnetic, and consequently, that there exists in the higher regions of the atmosphere an elastic fluid partaking of the properties of iron, or rather of magnetic steel . . . and . . . the light of the aurora borealis . . . is electric light solely, and that there is nothing of combustion in any of these phenomena.

John Dalton 1766-1844
English chemist and natural philosopher

A. W. Greely   1844–1935
Soldier, explorer, scientist, and author

Auroral Height

Before the early 1700's it was widely believed that the aurora occurred very near the earth. Some observers claimed to have seen the light in such positions between themselves and neighboring objects to demonstrate that the aurora descends to the very surface of the earth and may even be entirely confined to the lowest stratum. Others have reported it to be located among the clouds, so that its origin had to be placed at or below their level and therefore within some thousands of meters from of the earth's surface. Another argument for low auroral heights was that the rapidity of movement of some auroral forms seemed impossible to reconcile with any great distance.

Halley is generally credited with making the suggestion of a practical way to measure the height of aurora (210) (see page 52) by simultaneous observation from two separate locations. Chancellor Wolf of Halle University in Germany seems to have been the first to try to apply the technique, arriving at a height estimate of about 45 km (511). Mairan (see page 55) also began to try to apply the technique in 1726; he was aware of the difficulty of ensuring that both observers measured the same point of the aurora, as the effects of perspective can greatly modify the aurora's appearance. He decided to choose the lowest levels of the auroral arc in the vertical plane of the two stations and between 1726 and 1730 obtained heights of 500–1500 kilometers (306).

These results by Mairan were quite contradictory to the popular opinion at that time, though subsequent measurements throughout the eighteenth century generally confirmed the considerable height of the aurora. Most results, though, tended to be much lower than Mairan's estimates; Dalton (128) and Cavendish's (95) observations in England gave heights of 80–160 kilometers, while Loomis (293) in New England obtained 25–75 kilometers. The French Commission expedition to Bossekop made many height measurements (21), concluding that the aurora was 100–150 kilometers above the earth.

But uncertainties in ensuring that observers performing triangulation measurements were looking at the same point of the aurora continued to lead to a wide range of height estimates; in one instance, two groups of observers obtained 377 and 1313 kilometers for the same aurora (21). At the other extreme, many skilled observers continued to report very low auroras. Captain Parry at Fort Bowen in 1825 reported an aurora between his ship and the adjacent shore at an altitude of 670 feet, and observers on the Bossekop expedition sometimes claimed to see auroral rays between themselves and adjacent mountains (21). In England the Reverend Farquharson claimed (168) to have determined that auroras occur between 2000 and 5000 feet above the ground. Reports from Norway described how the aurora seemed to touch the earth itself, and on the highest mountains they produced an effect like a wind round the face of the traveler (149).

From the journal of Lieutenant Greely (203) at Fort Conger, Grinnel Land, on November 17, 1882, we have the following:

> The aurora of this morning was a very low one, and we are, I think the only party that could ever say we were in the midst of electric light . . . . at times the aurora could not have been more than 100 ft. from the earth . . . at times I raised my hand instinctively, expecting to bathe in the light.

It took the invention of the telephone and photography to finally put the question of auroral height on a firm scientific basis. The telephone allowed precise timing of auroral photographs and these photographs could then be compared with the background star field and allowed accurate positional determinations. Störmer (see page 135) accumulated some 40,000 such photographs between 1911 and 1944, and statistical analysis of these measurements (153) established an average auroral height of about 110 kilometers. Auroras rarely are lower than 90 kilometers, but some (red) aurora are often as high as 250 kilometers.

Dalton developed his auroral ideas in a strict mathematical formalization. From five mathematical propositions on auroras and five phenomenological statements, he went on to prove various other propositions, an example being *Proposition I*:

> The luminous beams of the aurora borealis are cylindrical and parallel to each other, at least over a moderate extent of the country.
> Proof: The beams must be parallel to each other, from Corol. to Prop. 2, and Corol. 2, Prop. 1, Sect. 1; and from Phenom. 1. Hence, and from Prop. 2, Sect. 1, and Phenom. 5, they are cylindrical.

Thomas Young   1773–1829
English physicist and egyptologist

At the same time, but independently of Dalton, an author writing under the pen name of Amanuensis described another magnetic theory of the aurora (16), which was probably influenced by Halley's earlier articles. Writing in *Mathematical, Geometrical and Philosophical Delights*, published in 1792, he supposed that magnetic effluvia are constantly issuing from the earth's magnetic pole; because of their ferruginous nature they fly off along magnetic meridians. He then conjectured that sulphurous vapors rising from the many volcanoes in the north mix with the magnetic effluvia, catch fire, and fulgurate, so that "a highly subtilized aerial nitre always enters into the compositioon of an aurora." He asked, "May not the luminosity be conveyed on the magnetic effluvia, as the electric on the iron wire?" Similar ideas were published in France in 1806 by Libes (282), who proposed that production of hydrogenous gases was less near the poles, so that sparks of electricity passing through the atmosphere caused production of nitrous acid and nitric acid, which would give birth to ruddy vapors. Their red color would vary according to relative concentrations of the nitrogenous compounds.

The English scientist, Thomas Young (who first introduced the idea of three primary colors in the theory of color vision), was also most interested in the aurora. In *Young's Natural Philosophy*, published in 1807 (519), he theorized that

> The matter of the aurora is of such a nature, as to be set on fire and when inflamed can shine with a faint light, and continue to be rare: For the stars may be seen through it. But who can pretend to assign its nature and properties, without great rashness? Chemistry supplies us with specimens of inflammable and phosphorean matter, that are almost innumerable; and nature has shut up in the bowels of the earth many others of different kinds, which science has not yet arrived at ... It would be better to attend diligently to the aurora; for perhaps the matter itself, at some time or other falling from the sky, may afford an opportunity of examination, or science may prepare something like it, by which it may be known; or its native place may be discovered in the superficies of the earth.

And resurfacing for the umpteenth time at the turn of the century, was the "new" theory of M. Monge (334) that ascribed the aurora to clouds illuminated by the sun's light which falls upon them after numerous reflections from other clouds placed at different distances in the heavens, even when the sun is considerably depressed below the horizon of the spectator.

The state of auroral physics at this time is summarized nicely by M. Haüy in a French text, *Natural Philosophy*, published in 1807 (215):

> ... it appears that every hypothesis has been tried to account for the aurora borealis. Among the different causes to which it has been attributed, we should be inclined to prefer that of electricity; but as yet this preference has no decided observation in its favour, and the uncertainty that still accompanies the phenomenon we have been considering, furnishes a new proof, that what we have known longest, is not always that which we know best.

Auroral Theories of 100 Years Ago

> Few phenomena have given rise to a greater number of hypotheses and theories as the polar aurora . . . it would be easy to fill a volume with the mere enumeration of these attempts at a theory, without discussing them.
> Alfred Angot
> *The Aurora Borealis*, 1896

Though somewhat exaggerated, Angot's sentiment (21) is well expressed. Even if consideration is limited to 100 years ago, we can come up with the following list of phenomena that various authors of the day associated with the aurora:

> weather (clouds, thunderstorms, wind, pressure, temperature gradient), seasons, earthquakes, volcanoes, moon phase, lunar halos, shooting stars, meteoric dust, telluric currents, magnetism, atmospheric electricity, sunspots, solar corona, Venus and Jupiter, zodiacal light, astronomical cycles.

And from these associations, various hypotheses, theories, and models were developed. The following tabulation lists just those ideas that were being discussed in the serious scientific literature in the mid-1800's to the late 1800's. For each entry the originator or a contemporary (1880) advocate of the idea is indicated.

Hypothesis or Theory or Model	Originator or Supporter
Reflected sunlight from ice particles	Descartes (139)
Reflected sunlight from clouds	Monge (77, 334)
Sulphurous vapors	Musschenbroek (342)
Nitrous gases	Libes (282)
Mixed mass of gaseous exhalations	Coates (90)
Burning of gases from putrefaction of animal and vegetable substances, ignited by falling stars	Parrot (363)
Luminous particles of earth's atmosphere	Euler (165)
Combustion of inflammable air	Kirwan (263)
Magnetic effluvia	Halley (210)
Luminous magnetic particles	Dalton (128)
Meteoric dust ignited by friction with atmosphere	Biot (55)
Atmospheric circulation patterns	Franklin (176); Rowell (399)
Electrified molecular circulation	Edlund (151)
Electric discharge between fine icy needles	Bradley (73)
Electric fluid in vacuum	Hawksbee (90); Canton (89)
Electric discharge in magnetic field	Lemström (279); Capron (90); de la Rive (134)
Electric discharge between earth's magnetic poles	Marsh (311)
Electric currents in aqueous vapor	Planté (370)
Condensation of vapors carrying latent electric fluid	Volta (486)
Thin clouds illuminated by free electricity flow	Holden (235)
Phosphoric electric light	Bertholen (53)
Phase of certain thunderstorms	Silberman (423)
Zodiacal light	Mairan (306)
Cosmic dust	Olmsted (358)
Cosmical particles	Humboldt (242)
Currents generated by compressed cosmic ether	Unterweger (477)
Solar particle streams	Becquerel (50)

> A theory has only the alternative of being right or wrong. A model has a third possibility: it may be right, but irrelevant.
> Manfred Eigen
> *The Physicist's Concept of Nature*, 1973

4
Severnoe Sijanie

But, where, O Nature, is thy law?
From the midnight lands comes up the dawn!
Is it not the icy seas that are flashing fire?

M. V. Lomonosov
Rhetoric, 1747

The earliest report of aurora that has been found in Russian records dates to the year 919 A.D., when an old chronicle reported, "This winter the sky was burning and fiery pillars were walking from Russia to Greece." In his book, *Russian History*, Tatiscev (460) interprets this as a reference to the aurora, which is corroborated by independent deductions that 918-922 A.D. was a maximum in the solar cycle. Arago's auroral catalog (27) also mentions "fiery spears" in the years 918 and 919, and the Arab traveler Ahmed ibn-Fadlan wrote a fanciful description (186) while visiting the capital of the King of the Bulgars (now the Russian village of Bulgary, 55°N) on May 11, 922 A.D.:

> ... and in the air I could hear loud voices and muffled noise. I raised my head, and a red cloud looking like fire was near me; from it came noise and voices, and in it one could see people and horses; and in the hand of those figures were swords, bows, and spears ... And here again came a similar cloud in which I also saw people with arms and spears, and the second cloud moved rapidly towards the first, like a squadron of cavalry attacks another. We became frightened and began to pray to God with obedience, but the natives of the country ridiculed us and wondered at our acts ... We asked the King about it and he told us that his forefathers said that the disciples of the Devil and those rejecting them were fighting every night ...

It is clear that the aurora was a common occurrence at the latitude of Bulgary at this time, as it was not feared by the natives of the country, who ridiculed the fears of their Arab visitors. It is interesting to note the remarkable similarity between the above description and that appearing in the *Book of Ezekiel* in *The Old Testament* (page 35).

The interpretation of the aurora as battles between heavenly armies was common in the Middle Ages, and such references can also be found in Russian history. During the great ice pack on the Dnepr River in November–December of 1066 there was a battle between Jaroslav and Sjvatopolk when, according to the Novgorod chronicle (454), "many righteous ones saw angels of God helping Jaroslav." On March 24–27, 1111, the Russians fought the Polovcy (western Slavs) on the Don and, despite being greatly outnumbered, won the battle. Captured Polovcy cried, "How could we fight you when there were angels riding over you in their bright armour and helping you?" Indeed, the Ipat'evskaya chronicle (453) reports that the Russians were advancing from the north, so did not notice what was presumably an auroral display in the sky behind them, whereas the Polovcy could see the whole northern sky. The aurora was strange to them and apparently frightened them and thereby influenced the outcome of the battle.

Further such references include the Tver' chronicle (453) which relates the appearance of a heavenly platoon during the battle at Lake Cudskoe in 1242, and in the Gustinskaja chronicle (453) of 1269 we read, "Great miracles appeared; the people saw armoured troops in the sky, divided into two squadrons and fighting each other." The Nikonovskaja chronicle (453) reports that "there were military regiments standing in the night air" in the year 1292.

The chroniclers of these times regarded it as important to report auroras, as it was believed by many that they accompanied or were followed by great misfortunes or troubles. But beginning in the fourteenth century, Russian monks were beginning to interpret symbolic indications of God's will in the aurora borealis. In the chronicles of the Voskresenskij monastery (454) is written,

In the year 1335 there was a vision . . . suddenly appeared a great light glowing in the west, like a dawn bending down from the sky . . . and Father Afonasij said to prince Semenovic, 'On this spot God wants to radiate a divine bliss.'

In 1397 the monk Kirill was praying in the Simonov monastery when he heard a

. . . voice speaking: 'O Kirill, go to the White Lake and you will obtain good peace.' Having opened the window, he saw a great light shed from the sky and radiating the Northern countries, moving towards the White Lake as if pointing with its finger the destined place . . . and he spent all night rejoicing and praising God.

The monastery at White Lake founded by Kirill about 1400 A.D. on the argument of Divine justification in the form of an aurora.

Soon after, he founded a monastery at White Lake and later the regional town of Kirillov (near Petersburg). However, the monastery records tell us that the Simonov monks had previously investigated the White Lake as a good monastery site, as it was at the crossroads of many important trade routes. But the local Novgorod leaders were resisting attempts at colonization of the area from the Petersburg and Moscow regions. Kirill must have decided that a divine justification of this earthly business was needed from above, and an appropriate sign was found in the aurora borealis. His scheme was very successful, and the Kirillo-Belozerskij monastery became one of the richest and most influential in Russia (454).

Mairan's auroral catalog reports a great aurora over Europe on August 27–28, 1581, and this aurora is mentioned in many Russian chronicles. An old semiblind blacksmith in Pskov saw a bright light and pillar which he interpreted as the Blessed Mother and many saints, a view endorsed by the Pskov clergy (454). But a scribe in the office of Stefan Batorij (King of Lithuania) wrote a curiously dispassionate description (329) (for the times):

On the night of August 27 there were seen signs in the sky, resembling pillars like two armies of cavalry, and others resembling crosses. But there is no miracle here, and is likely to be some game of nature, an evaporation or the like.

This remark is an exception to the usual descriptions of battles between divine armies in the sky, which were common in contemporary Russian, Polish, and German chronicles. But more surprising still is the opinion of an anonymous scribe from the Cudova monastery (454), written sometime between 1586 and 1600:

. . . About the pillars moving in the air; they appear when the sun slips under the earth to the west, and some rays extend to the air in the north. There the air lights like an arch or like a folded tablecloth, and from that airy light a shining appears on the ocean. The ocean is uncalm with waves, and so the shining moves back to the air in the form of bright pillars. As proof, you can take a vessel of water to the window and move the vessel and vibrate the water in the sun's light. Immediately on the ceiling are the sun's rays flickering from the water.

Such an optical hypothesis is remarkable in a period of earnest belief in signs and miracles. The unknown scribe is aware of his audacity for he concludes with

And, dear reader, having read this brief creation of my sinful hands and research of my rough mind, do not blame my narrow-mindedness—and where it is wrong, feel free to correct.

Such natural explanations for the aurora did not develop in Russia as the seventeenth century progressed, as the aurora practically disappeared from the skies at this time. Its absence resulted from the long-lasting and substantial decrease in sunspots between 1645 and 1715 (the Maunder minimum). Auroras were even rare in Norway and Iceland, and Kochen, the Swedish envoy to Moscow, described a rare aurora seen in 1688 as new and unusual to him (266).

Peter I (the Great) 1672–1725
Czar of Russia (1682–1725)

The long absence of auroras from Russian skies ended in a spectacular fashion on March 17, 1716. The great aurora of that night was seen over vast areas, and reports came from Poland, the Ukraine, Spain, Portugal, Italy, France, England, Switzerland, Holland, Austria, Germany, Sweden, Norway, and sites in North America. This was the aurora that Halley observed in London; and in Petersburg, Peter the Great wrote in his diary (367), "On March 6, [old system] at 9 p.m., there was a terrible sign in the sky; the night was very bright, as if there was a full moon." The astronomer Brjus wrote to Peter about this aurora (454) expressing his surprise at the "continuity of this alternate burning, because when meteors appear they quickly burn and disappear."

The historian Tatiscev reported the following description (460) by peasants from Krasnoe Selo:

. . . there was seen over the castle of St. Petersburg an impression as if made by letters in the hands of a printer—the Lord in bright radiance—there was an eagle soaring as if for battle . . . and three dark clouds moving to fight the eagle . . .

However, he comments that though he was standing with them, he could see nothing of that kind. And after a century long silence the chronicles of the Russian monasteries began again to speak of the aurora, and lists were compiled of their appearances.

Peter the Great again witnessed a spectacular aurora (367) while at sea near Astraxan on October 3, 1722 [old system] (46°N); it was seen in the southern sky:

. . . a great fire . . . began to move rapidly to the west with the same width with which it began, and from where it left it no longer was . . .

The extreme southern visibility of this display is consistent with reports of auroras on the same night from Paris and Berlin and also from Lynn, New York (42.5°N)(21).

The first attempts at scientific explanation of the aurora in Russia

date back to the 1720's and the academicians Meyer, Kraft, and Geinzius at the St. Petersburg Academy. Meyer suggested (328) that as the sun
gradually moves away in the autumn, a "chill" grows on the northern
hemisphere; a "great fire" is then taken off the warmer parts of the
earth, and the vapors catch fire along the boundary of cold and warm
air. In the 1730's Kraft (268) took issue with current suggestions that the
aurora was a light reflected from the Hekla volcano by moving and
broken ice in the sea. He said he had never seen aurora during the breaking of ice on Ladozskoe Lake and argued "how could Hekla mountain
with all its fiery coals have such a great power over ice and snow in the
Northern country, as to have this ice flooded by light so cruelly that the
reflection from it could be seen all over Europe?" The aurora was of
prime interest to Kraft, who organized a group of amateur observers in
St. Petersburg; he tried to measure the height of the phenomenon on two
occasions, obtaining unrealistically high results of 578 and 676 miles.

Wolfgang Georg Kraft 1701-1754
Russian physicist

In 1740, Geinzius wrote an extensive review (193) of Mairan's textbook
Traité Physique et Historique de l'Aurore Boréale (published in Paris in
1733) and expressed agreement with Mairan's views that related aurora
to the zodiacal light. In fact he was so keen to support the theory that in
reporting on the famous comet of 1744 he confused the spectacular
comet tail with the aurora and stated that it appeared simultaneously
with strong zodiacal light.

Gottfried Geinzius 1709-1769
Russian physicist

This period saw the discovery by Hiorter (233) and Celsius (96) (in
Sweden) of the association between aurora and magnetic disturbances.
But possibly this association was already known in Russia; around 1716,
Russian sailors watching the magnetic needle on long voyages to Spitz-
bergen mentioned that "during pazori [aurora], the needle goes crazy"
(454).

The old Russian terms (454) for the aurora were *pazori* and *spoloxi*. The
word *pazori* is very old philologically and comes from *pa*, meaning
similarity (but not a complete likeness), and *zori*, meaning dawn. Thus
the literal translation is "not real dawn." Spoloxi comes from the archaic
Russian word *spolox*, or *upolox*, meaning alarm bell, and *polosit*, meaning to sound an alarm (516). Various other terms described specific
auroral types: the beginning of pazori, a pale white light in the northern
sky, was called *otbel*; when otbel reddened it was called *zorniki*; milky
strips in the sky were called *luci*; and if luci reddened they were *stolby*.
When actively playing, stolby were said to be accompanied by a roaring
sound like thunder and then called spoloxi.

In old Russia the attitudes toward the aurora borealis were different in
the northern and southern parts of the country. Some interpretations in
the early 1700's by northern people were mystical: "the angels are playing" or "as a sermon to the people so everyone can see with his own eyes
how, from the mouth of the righteous ones, a prayer is seen as fiery
pillars rising in the sky." But, in general, the beliefs of northern Russian
sailors and of the people of Siberia were associated with meteorological
phenomena, whereas in the south, where auroras were much rarer, connections were made with important events such as wars and famine.

It is possible that the old Russian name for the month of August was
derived from the aurora. The "white nights" end in northern Russia
around the end of July, when short periods of darkness return between
the evening and the morning dawn. Appearances of the aurora during
this brief darkness are thought to have given rise to the old names *zarev*

Mikhail Vasil'evich Lomonosov
1711-1765
Regarded by many historians as the
founder of Russian science

and *zornicnik* — the beginning of auroras after the summer break. There is a special date in the traditional folk calendar of northern Russia, *Felka-Zarevnica*. The word *zarevnica* meant the peak of "polar dawns," or aurora; there is a maximum of occurrence in October, and the day of Felka comes in the beginning of October. The name *Felka* seems to result from an association with a legend told in *Cet'ja-Mineja,* in which Felka is ordered by a Prince's decree to be thrown to wild beasts to be eaten up. Instead, with the words "in the name of Christ I shall be baptized on my last day," she jumped into a lake (454).

> The Prince began to cry because such a beauty was to be had by seals. But the seals, seen shining like fiery light, dropped dead. And after that there was a fiery cloud about her, so that no beasts could approach her, nor could the people see her nudity.

The mid-eighteenth century in Russia brings us to the work of Mikhail Vasil'evich Lomonosov (1711–1765), credited by Russian historians as the founder of Russian science. Born of peasant stock in the arctic north, the young Lomonosov rejected his father's business of fishing and trading and at 19 he misrepresented himself as the son of a priest to enter theological school in Moscow. His aim was to learn Latin so that he could pursue his interests in natural science. He showed outstanding promise and was sent to Germany to study chemistry under Wolf and Henkel, returning to St. Petersburg Academy in 1741. Here a personal animosity with the director, often leading to physical violence, hindered his professional career. He wrote on history, languages, and science, though much of his work remained obscure until 1903 when Menshutkin began to research his writings. His papers included an active correspondence with Euler on scientific matters (289).

From early childhood, Lomonosov was especially curious about the aurora, which was a common event in his hometown but was still mystifying and demanded a natural explanation. Lomonosov lived at the time of Benjamin Franklin's famous electrical experiments and proposed that the auroral light was similar in nature and origin to the luminescence present in electrical discharges. According to his theory, first presented (288) in a speech to the St. Petersburg Academy in 1753, atmospheric electricity is caused by friction of frozen vapor particles of the "upper atmosphere" when they descend as more dense ones into the "lower atmosphere" through an intermediate level of "middle atmosphere." Simultaneously, particles of the warmer and therefore lighter lower atmosphere move upward.

Lomonosov conducted many electrical experiments to examine the luminescence of gases under low pressures when subjected to electric current and repeated Franklin's experiments on atmospheric electricity. He thought that the auroras extending beyond the earth's atmosphere were white, whereas "on the very surface of the atmosphere, the radiances and pillars are born in the ether by the movements of various vapors of different colors."

Lomonosov's auroral theory (289), including atmospheric circulation of air masses, was remarkably similar to Franklin's ideas on the aurora, though Lomonosov was sensitive to such comparisons.

> In my theory of electric force in the air, I do not owe anything to Mr. Franklin. I deduced the cause of it as the result of lowering of the cold upper atmosphere at the start of bitter frosts; that is, from circumstances unknown in Franklin's native Philadelphia.

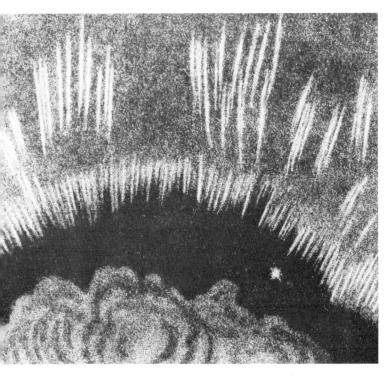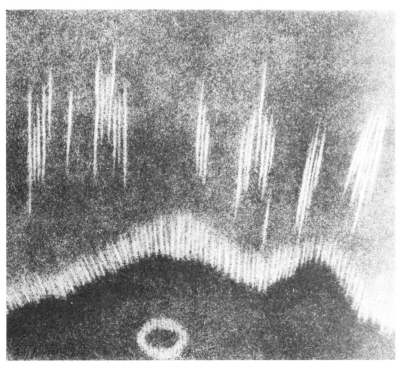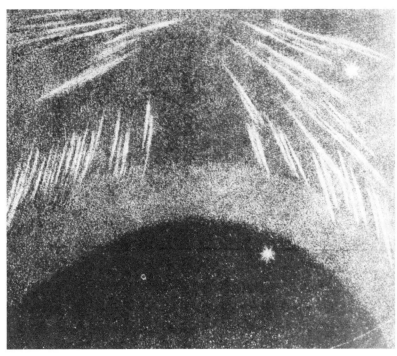

Impressions from one of the copper plates engraved to illustrate Lomonosov's planned three-volume treatise *Research into the Origin of the Aurora Borealis.*

F. P. Wrangel 1796–1870
Russian arctic explorer

In 1763, Lomonosov was involved in a project to organize an expedition to look for a northeastern passage to Siberia and returned to the question of the aurora borealis. During the project he painstakingly tried to prove the existence of a great northern unfrozen patch of water in the midst of a surrounding ice field and again developed his theory about rising and falling streams of air. In his opinion the rising streams in the north would only be possible because the ocean remains free of ice there in some places.

In the last years of his life, Lomonosov decided to write a large monograph in three parts under the title *Research Into the Origin of the Aurora Borealis*. Only the outline and first two chapters of part 1 (which was to describe Lomonosov's observations, reports of other authors, and the oral traditions of Siberians) has reached us. Part 2 was to present the *Theory of Electric Force*, and part 3 was to deal with *Manifestations of the Aurora Borealis*. To supplement the work, the Academy of Sciences ordered and engraved 11 copper plates showing 48 auroral drawings by Lomonosov, but the dates and explanations were lost so the planned publication never took place. Impressions made directly from the plates have survived in the Archives of the Academy in Leningrad and were

About 1600 A.D. a scribe in a Russian monastery explained auroral structure as resulting from reflection of the sun's light from the uneven waves of the ocean. This contemporary painting by Stephen Hamilton is of the aurora off the coast of Labrador.

published as a supplement to volume 3 of the *Complete Collection of Papers of Lomonosov*, printed as 10 volumes by the USSR Academy of Sciences between 1951 and 1959.

After Lomonosov's death, research on the aurora declined in Russia; the only interesting work that emerged from the nineteenth century was that of Wrangel (1824) (516), who was the first to claim an enhancement of aurora near the shoreline of northern Siberia. Further observations on the possible effect of the coastline on the occurrence and alignment of aurora were carried out from the vessel *Zarya*, captained by Toll', on the 1900–1902 expedition to look for a hypothetical Sannikov's land in the Arctic Ocean. The careful data that were gathered were not analyzed until 1963 when Nadubovich reported (343) that the coastline effect was confirmed.

Since the Second Polar Year in 1930–1931, systematic auroral observations have been carried out in Russia, and auroral physics continues to be an important part of the upper atmosphere and space research efforts in the USSR today.

E. V. Toll' **1858-1903**
Russian arctic explorer

Aurora in Alaska. Painting by F. Jaikema.

Aurora observed at Turin, Italy, on February 29, 1780.

5
A Century of Observations

Give me to learn each secret cause;
Let number's, figure's motion's laws
Revealed before me stand;
These to great Nature's scene apply,
And round the Globe, and through the sky,
Disclose her working hand.

Mark Akenside
Hymn to Science, 1779

It is not surprising that, after a
long period of searching and erring,
some of the concepts and ideas in
human thinking should have come gradually closer to the fundamental laws
of this world.

Victor Weisskopf
Knowledge and Wonder, 1966

Sir John Franklin 1786-1847
Ill-fated English arctic explorer

Elias Loomis 1811-1889
Professor of Natural Philosophy and
Astronomy at Yale College

Herman Fritz 1830-1893
German physicist

Although the eighteenth century had seen the beginning of the realization of the electric and magnetic nature of the aurora, these ideas were striving for acceptance over the more bizarre suggestions of volcanic sulphurous compounds and aerial nitrous gases and acids. This time in history was the beginning of a period of great geographic exploration, and adventurers from many countries endured great trials and hardships on polar expeditions. These expeditions resulted in some of the first extensive synoptic auroral observations, often with electric and magnetic instrumentation. In the early nineteenth century, numerous scientific arctic expeditions were organized by national scientific bodies, auroral observations being one of the major motivations.

The nineteenth century saw the realization that there is an auroral zone, a region of maximum auroral occurrence. It was known, of course, that auroras become more frequent as one traveled north from temperate latitudes, but Captain John Franklin, the ill-fated arctic explorer, seems to have been the first to comment that the frequency did not increase all the way to the poles (177). In the narrative of his 1819–1822 journeys he states that

> ... were I to venture an opinion as to its probable situation [the region of maximum auroral frequency], I should say between the latitudes of 64° and 65° north, or about the position of Fort Enterprise, because the coruscations were as often seen there in the southern as in the northern parts of the sky, and I should consider that latitude the most favourable in this part of the globe for making good observations of this interesting phenomenon.

Franklin's tragic expedition of 1845 to try to discover the Northwest Passage (on which the whole party of 129 officers and men perished) led to a series of search and rescue expeditions over the next decade. Many of the memoirs written by the leaders of these expeditions described the beauty of the aurora, and the popularity of such books led to an increased awareness of the aurora by the general public.

In 1833 the German geographer Muncke used accounts from many polar explorers to infer that the frequency of auroral occurrence did not increase all the way to the pole (341). Kämtz may have been the first to chart curves of equal auroral frequency (318), and subsequently, both Loomis (292) and Fritz (179) plotted the geographical distribution and mapped out zones of equal frequency. Loomis, a professor of natural philosophy at Yale, supplemented historical data with records kept in New England to prepare the first map of the auroral zone, which he published in 1860. Fritz searched world archives with unremitting energy to collect further European records and extended Loomis' work. In 1874, while a young professor at the Polytechnic Institute in Zurich (which was later to nurture the young Albert Einstein), he published his isochasms, lines of equal average frequency of auroral visibility. (The term arose from *chasmata*, Aristotle's original term for a particular type of aurora.)

It was evident, thought Fritz (179), that auroral frequency varied in consonance with the waxing and waning of sunspots. What was most puzzling was that the center of the ovals of equal auroral frequency was located neither at the north pole nor at the magnetic pole but at a point northwest of Greenland, some 800 miles from the magnetic pole. This apparently marked the axis of the earth's magnetic field at high altitudes, and came to be known as the geomagnetic pole. The differences from the magnetic pole were attributed to local magnetic irregularities within the earth.

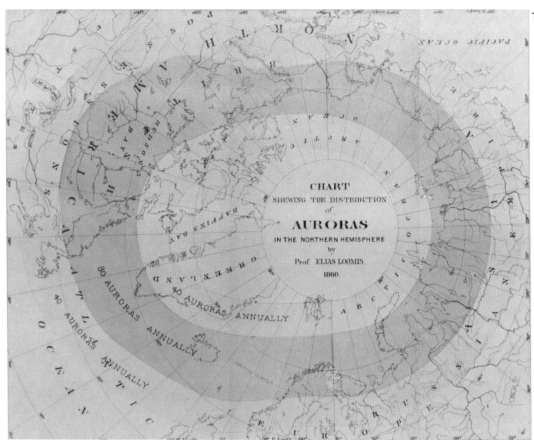

Isochasms (lines of equal auroral frequency) determined by Fritz in 1881. At the region of maximum frequency, Fritz indicates that auroras are seen on more than 100 nights per year, whereas at mid-latitude he shows the frequency decreases to 0.1 (one aurora every 10 years).

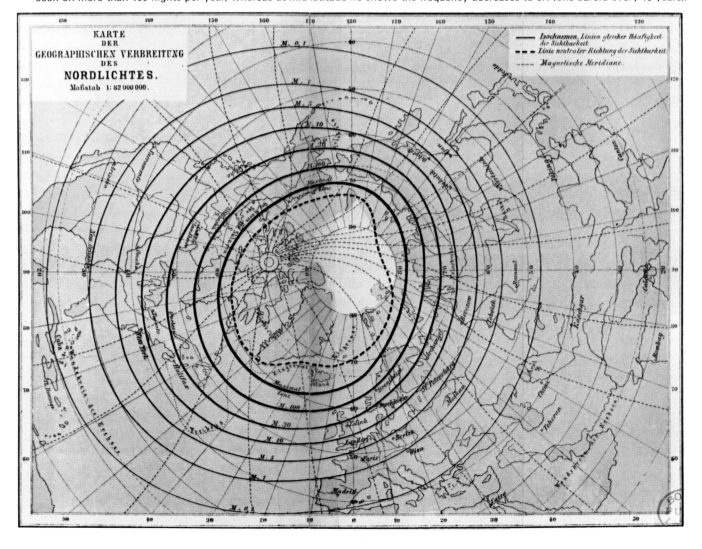

The great French scientists Gay Lussac and Biot were interested in problems concerning the earth's magnetism and the upper atmosphere. Here they used a balloon to reach high altitudes (4000 meters) to sample the atmosphere and to measure whether the strength of the earth's magnetic field decreased with height. Both these measurements were relevant to auroral theories of the period.

Jean Baptiste Biot   1774-1862
French mathematician and philosopher

Count Alessandro Volta   1745-1827
Italian physicist

D. F. J. Arago   1786-1853
French astronomer

Fritz was among the first to cite evidence that auroral displays occurred simultaneously at both ends of the magnetic field line. Derivations of isochasms for the southern hemisphere had to wait for the development of scientific stations in Antarctica and were published for the first time in 1939 (496).

Various theories of the origin of the aurora continued to emerge throughout this period. Biot had studied the light of the aurora with a polarimeter in the Shetlands and in Scotland in 1817 and could find no trace of polarization (55). This proved beyond a doubt that the aurora was self luminous and not the result of reflection or refraction (as are rainbows or halos). Biot suggested that this light could come from ferruginous particles in the air from volcanic eruptions. This was essentially Dalton's idea (128) and it was revived again in 1840 by von Baumhauer in Utrecht, who suggested a cosmic origin for the luminous particles.

In Italy, Volta had suggested (486) (about 1815) that when water vaporized at the earth's surface it became enriched with electric fluid; when these vapors condense at high altitudes and high latitudes because of the cold there, their ability to store electric fluid is lost, and the excess electricity is seen as the aurora. He was aware of certain problems in getting such vapors to the heights at which auroras occurred and was also investigating alternative nonelectrical theories for the aurora.

The *Smithsonian Contribution to Knowledge* volume of 1856 contained a careful and incisive discussion of the current understanding of aurora by Olmstead (358), a professor at Yale. After studying "a greater amount of facts, than, so far as I know, any other person has taken the trouble to accumulate," Olmsted arrived at 11 "laws of the Aurora Borealis." From these laws he arrived at his conviction that the aurora was not in the nature of an electrical discharge but was "cosmical in origin, the matter of which it is composed being derived from the

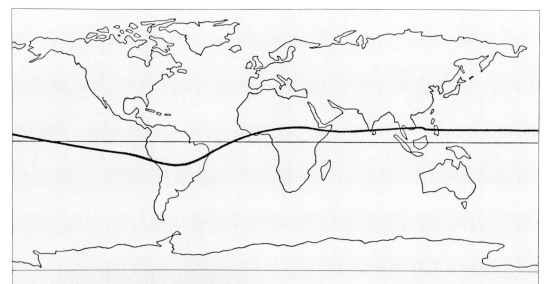

Map of the world showing how the geomagnetic equator deviates from the geographic equator, as auroras roughly follow parallels of geomagnetic latitude. It may be seen that a great aurora borealis will extend much further south over North America than over Central Europe and Asia. Similarly, auroras are much more likely to be seen in southern Australia than in Argentina.

Great Auroras

A great aurora seen over Paris, France, on April 15, 1869, at 10:30 P.M., by M. Silberman.

The polar lights sometimes leave the Arctic and Antarctic and are observed at much lower latitudes. In fact, the aurora may have been seen from the equator on rare occasions. Such events follow gigantic solar activity, are accompanied by worldwide magnetic disturbances, and might only occur once or twice in a century.

There are a few biblical records of auroras (see Chapter 1), and the northern lights reached latitudes of 30°–35°N on these occasions. There is a tradition of a shower of blood before the birth of Mohammed in 570 A.D., which is consistent with mentions of auroral activity in the second year of the Byzantine Emperor Justin II, in 566 A.D. (65). Auroras seen rarely from Ceylon (7°N) were called Buddha's lights (429).

Probably the greatest aurora of all for which we have reliable records was that of September 1, 1859. This was the occasion that Carrington first observed a solar flare (92) (page 81), and an ensuing aurora was seen in the zenith in Puerto Rico (262) (18°N). On February 4, 1872, a great aurora was seen from Bombay (19°N) and throughout Egypt (64). An aurora of September 25, 1909, was said to be seen at Batavia (now Djakarta, 6°S) and Singapore (1°N), though the reference seems somewhat doubtful (64). However, *aurora australis* was definitely seen at Samoa (19) (14°S) on May 15, 1921.

The above-quoted latitudes are geographic, but the aurora is controlled by the earth's magnetic field and hence is dependent on geomagnetic latitude. The geomagnetic equator deviates from the geographic equator by as much as 17° at longitudes of South America (see map above). Consequently, the northern lights have their best chance of reaching the geographic equator in that region. Thus aurora seen over Mexico (20°N) on February 11, 1958, was actually 33° from the geomagnetic equator, whereas those seen from Bombay (19°N) in 1872 were only 10° from the geomagnetic equator.

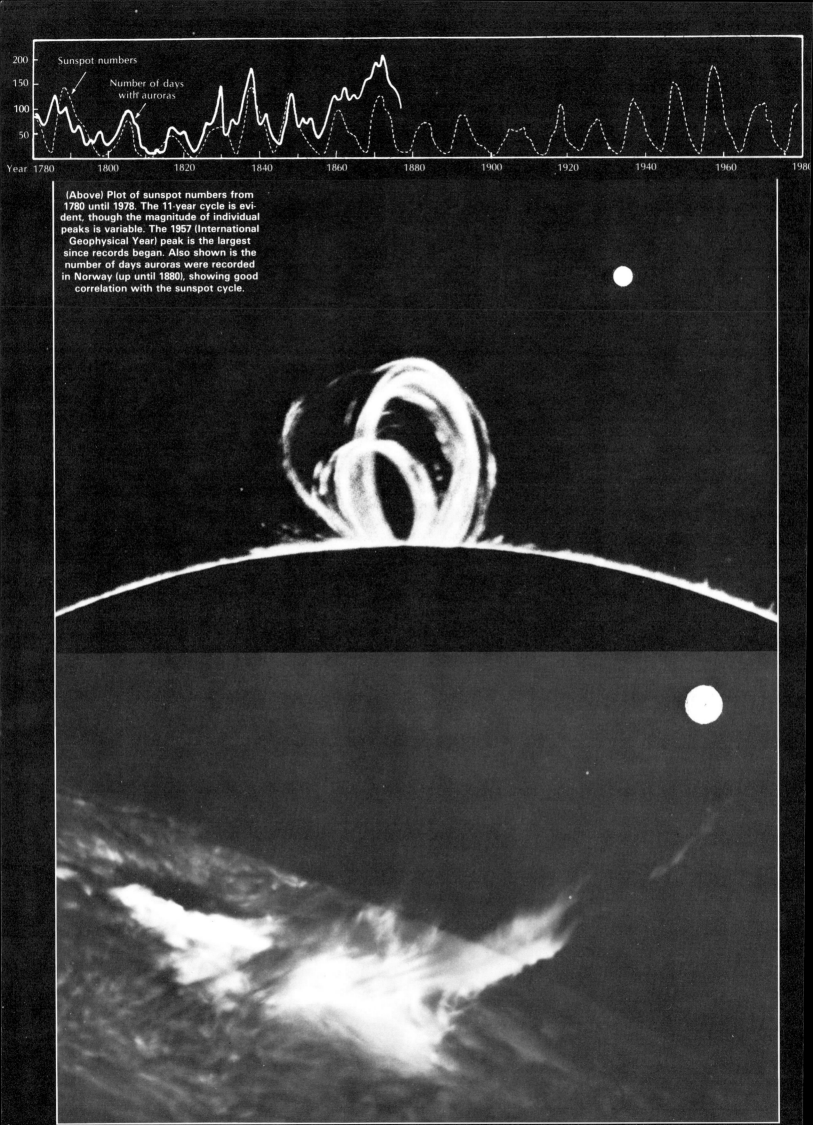

(Above) Plot of sunspot numbers from 1780 until 1978. The 11-year cycle is evident, though the magnitude of individual peaks is variable. The 1957 (International Geophysical Year) peak is the largest since records began. Also shown is the number of days auroras were recorded in Norway (up until 1880), showing good correlation with the sunspot cycle.

planetary spaces." Omsted had expressed these ideas in lectures to students at Yale as early as 1835, but in scientific circles they seemed to fall on deaf ears, as electrical theories were still very much in vogue. In support of his cosmical theory, Olmsted cited the great extent of aurora, their local time similarity at widely different longitudes, their great velocity (too great for any terrestrial matter), and their secular periodicity. Olmsted is less certain when considering the nature of the auroral vapor itself and also leaned toward Dalton's idea that it is "ferruginous," which he defined as "iron in a state of extreme diffusion." Similar ideas of "myriads of meteoric bodies travelling separately or in systems around the sun...consumed in thousands daily by our own atmosphere" were advanced by Proctor in *Half Hour Recreations in Popular Science*, published in Boston in 1874 (378).

An important clue to the origin of the aurora was sighted on September 1, 1859, although its discoverer was unaware of its full significance. Richard Carrington, an English astronomer, was drawing sunspot groups from a projected image of the sun (92).

> ...when within the area of the great north group...two patches of intensely bright and white light broke out...My first impression was that by some chance a ray of light had penetrated a hole in the screen attached to the object-glass, for the brilliancy was fully equal to that of direct sunlight, but...I was an unprepared witness to a very different affair. I thereupon noted down the time by the chronometer, and, seeing the outburst to be very rapidly on the increase, and being somewhat flurried by the surprise, I hastily ran to call someone to witness the exhibition with me, and on returning within 60 seconds, was mortified to find that it was already much changed and enfeebled.

But Carrington could not be accused of exaggeration, as some miles away a Mr. Hodgson had also been engaged in observing the same sunspots and saw the same dazzling brilliancy bursting into sight from the edge of the sunspot (234).

They had witnessed a solar flare—an event of awesome magnitude that envelops millions of cubic miles of space with the heat of a nuclear explosion, all within a few minutes. At the moment Carrington saw the flare, the Kew Observatory (London) magnetograms were disturbed, and 18 hours later one of the strongest magnetic storms ever recorded broke out. Auroras spread over much of the world and were seen as far south as Puerto Rico (261) (see page 79).

Carrington was cautious not to draw too definite a connection between his solar flare and the ensuing auroral displays and felt additional examples would be needed before a causal relationship could be established..."One swallow does not make a summer."

The spectrum (component colors) of the auroral light was studied for the first time by Angström (22) in 1866–1867, who showed it was essentially a spectrum of lines and therefore produced by luminous gases and not by solid or liquid incandescent particles. Even though the observed spectral lines did not correspond with any known spectrum of the atmospheric gases, it put to rest the theories of glowing ferruginous particles and liquid nitrous or nitric acids as being the constituents of the auroral matter, as such sources would give continuous or band spectra. In 1871, Zölner expressed the belief (521) that the auroral spectrum differed from that of the atmospheric gases only because it was a spectrum of another form in the atmosphere which as yet was incapable of artificial demonstration.

Richard Proctor   1837–1888
English astronomer and popular science writer

Although Carrington (1826–1875) was a prominent astronomer in his day and a Fellow of the Royal Society and the Royal Astronomical Society (he was secretary of the latter for 5 years), an extensive search failed to uncover a photograph of him.

(Opposite page, middle) A solar flare photographed in Hα light. The eruption extends some 10,000 kilometers into space. The white disc shows the approximate size of the earth.
(Opposite page, bottom) A gigantic solar prominence, photographed in Hα light, extending some 60,000 kilometers into space. The white disc shows the approximate size of the earth.

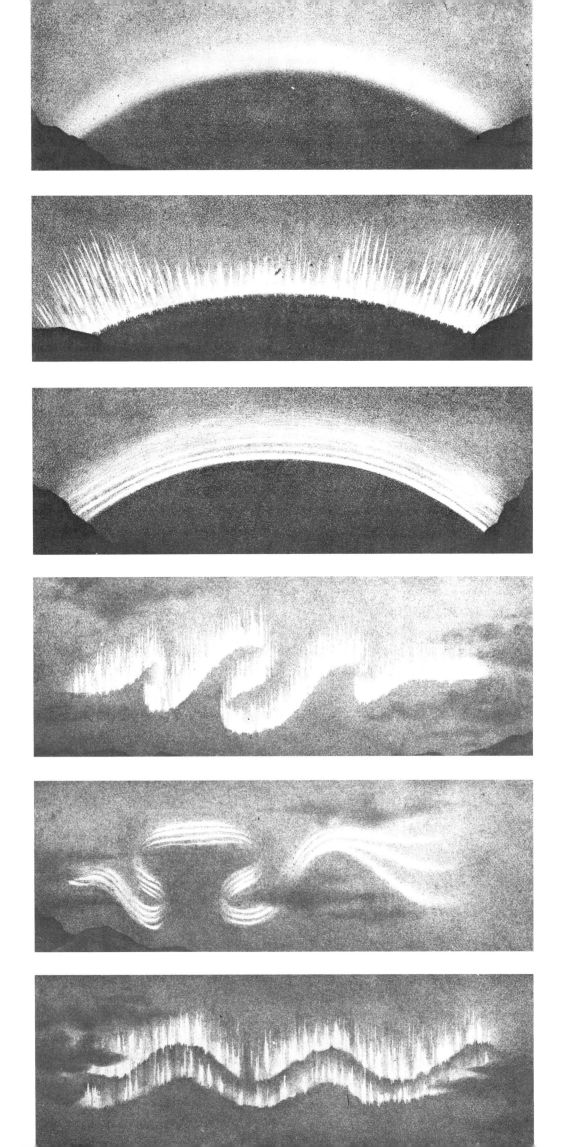

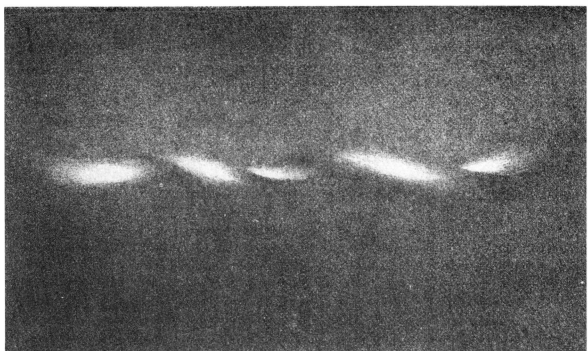

These plates show sketches of the aurora made from the Austrian polar station Jan Mayen during the First Polar Year, 1882–1883. Before the development of auroral photography, sketches such as these were the only permanent records of auroras that scientists could obtain for study.

**Sophus Tromholt 1851–1896
Norwegian teacher and
scientist**

Modern auroral physics is seen as having started with the first international cooperative study of polar regions in the First Polar Year (1881–1882), which resulted in voluminous national reports published by participating countries. But by far the most astute and penetrating treatment of auroral phenomena resulting from the First Polar Year is not to be found in any national report but has been condemned to relative obscurity because it constitutes just one chapter of a two-volume travelog (472) entitled *Under the Rays of the Aurora Borealis* by Sophus Tromholt, published in 1885. Tromholt was an official First Polar Year observer stationed at Kautokeino, in Norwegian Lapland; he became fascinated with the Lapp culture, and his book mainly concerns them.

Tromholt must be the most underestimated auroral scientist of all time. His irreverent and racy prose is full of marvelous insight and outrageous criticism, and the following selections are a few examples that illustrate his genius:

> But, although Science must admit its incompetence to give a reply to *the* one demand of the crowd [the cause of the aurora], consolation may be found in the circumstance that Science can in any case say what the aurora is *not*, and recommend the manufacturers of plausible auroral theories to find more profitable fields for their imagination. In answer to popular hypothesis, modern Science can declare with confidence that the aurora borealis is *not* sunshine reflected from the icefields of the polar regions, neither the reflection of sunrays in ice crystals suspended in the upper strata of the air, nor the reflection of a molten fluid in the earth's interior . . . etc., etc.

> Not only can Science reject such childish explanations . . . but can state with confidence that the aurora borealis is of an electrical nature and closely related to the terrestrial magnetic forces of the earth.

> The Scientific theories . . . are more abstruse than the popular ones, but equally fail . . . They are generally so untenable that their evolution can only be explained by assuming that the inventors have only studied the phenomenon in southern latitudes, where the aurora never attains the intense form which is its characteristic in the north. Having viewed a few aurorae, and been struck with the wonderful spectacle, a theory has at once been constructed for the entire phenomenon. And what one theorist has observed he has supplemented with such observations alone as have supported his own views—and the problem is solved. It is, however, only a pity that the real aurora is so very different from the theoretical. One might just as well form a theory for explaining the formation of glaciers from seeing an icicle, or construct the grammar of a language from the knowledge of a dozen words.

> . . . One need only open any one of the popular scientific works . . . to find it swarming with inaccuracies, and presenting the most extraordinary illustrations of the same, wholly the product of the draughtsman's imagination. And without any knowledge of the phenomenon, one compiler copies from another, the correct with the incorrect, and in the course of time the auroral literature has become so inundated with what might be called 'traditional falsehoods,' that it would be a Herculean task to clean these Augean stables.

**Tromholt at his auroral station
at Kautokeino, Lapland, 1882.**

After a beautiful description of the physical appearance of the aurora, Tromholt continues

Such is the Aurora Borealis in its utmost grandeur. No colour, no brush can picture its magnificence, no words describe its sublime beauty.

And below we wretched Children of Man stand and boast of our knowledge and our progress, conceited by possessing the mind by which we have defrauded Nature of some of her secrets; there we stand gazing at that great problem which Nature writes in flaming cypher on the dark winter sky; there we stand, lost in wonder, obliged to confess — that we really know nothing.

But despite his disclaimers, Tromholt proceeds to give an accurate explanation of the form of the aurora and draws a model of a ring of light emission around the magnetic pole, noting that

. . . on account of the great circumference of the earth, in proportion to the height of the aurora, only a small proportion of such a ring would be visible at one time, and each observer only see his own portion, the situation of which in relation to *his* horizon and the zenith will depend on *his* position in relation to the auroral ring . . . and further, that the ring is not always a single one, but consists generally of several, each having its centre in various points of the magnetic axis, and that the ring is seldom perfect, but generally broken, and with many deviations from a symmetrical configuration.

He then gives accurate observations that auroral arcs align along parallels of magnetic latitude, that rays or streamers align along the magnetic field, and that the coronal form is simply an effect of perspective. He comments on exaggerated notions of the strength of the auroral light; while admitting to momentary intense brightenings illuminating the landscape, he says it is a fable that northern people use auroral light to travel or to hunt and it is of no practical value to them. His own attempts at auroral photography with the most sensitive dry plates and 4-to 7-minute exposures never resulted in the faintest trace of a negative. (See page 253.)

Tromholt described how intense aurora "always cause severe derangement of the telegraphs . . . sometimes the whole telegraph system of a country may be rendered inoperative . . . if the disturbing current itself is not used for dispatching messages, which indeed may be done." (Early telephone technicians in Boston around the turn of the century also found that they could transmit to New Hampshire during a strong aurora without the necessity of a battery.)

The two most reasonable auroral theories of the day are discussed by Tromholt. He categorizes the cosmic theory whereby the earth enters clouds of ferric dust in space, which would gather in auroral configurations under the influence of terrestrial magnetism, as having no other adherents than its originators. He felt that the electrical theory deserved more adherence and described Professor Edlund's theory (151) (published in Stockholm in 1878) whereby the rotating magnetic field of the earth generated a voltage between the equator and the pole, resulting in a dissipative current flowing through the high atmosphere. Discharges would occur more easily in the direction of the magnetic field. Consequently, voltage would build up near the equator, where the magnetic field is horizontal, until finally lightning would occur when the voltage caused the lower atmosphere to break down. Nearer the poles, however, the magnetic field inclines and facilitates an even discharge towards the earth, generating the aurora.

Erik Edlund 1819-1888
Professor of Physics, University of Stockholm

Aurora seen from Bossekop, Norway, on January 21, 1839, at 6:00 P.M.

In 1881, Tromholt wrote a significant paper (470) that seems to have been largely overlooked. It appeared in the *Danish Meteorological Year, 1880* (published in 1882), and in it, Tromholt reported that the zone of aurora was displaced within the course of 24 hours, with the time of auroral maximum being later as one moves further poleward. He explained that this was due to the geometry of the auroral region, and so effectively discovered the "auroral oval" (see discussion on page 175), though he never called it that. He also noted the overall expansion of the oval with the solar cycle, the aurora locating further equatorward during years of peak sunspot activity.

Tromholt certainly did not feel the problem was solved, and his concluding remarks sum up the state of the auroral physics around the mid-1880's:

> Will Man ever decipher the characters which the aurora borealis draws in fire in the sky? Will his eye ever penetrate the mysteries of Creation which are hidden behind this dazzling drapery of colour and light? Who will venture to say? Only the Future knows the reply. But nevertheless the student toils yard by yard along the fatiguing road of research in the hope—maybe vainly—of some day reaching the much-coveted goal.

The Swedish government also sent an expedition to the Arctic for the First Polar Year, and their senior observer, Carlheim-Gyllensköld, carried out a meticulous program of visual observations (see below) from Cap Thordsen in Spitzbergen during the 1882-1883 winter (91). His analysis also revealed a systematic movement of auroras in latitude as a function of local time. He thus confirmed Tromholt's result but was not the discoverer of this effect, as has been claimed (15).

A number of polar expeditions of the 1880's claimed to have shown that as the arctic sea alternately froze in the winter and thawed in the summer the aurora followed the limits of sea ice in its latitudinal excursions.

**Vilhelm Carlheim-Gyllensköld   1859–1917
Professor of Astronomy, University of Stockholm**

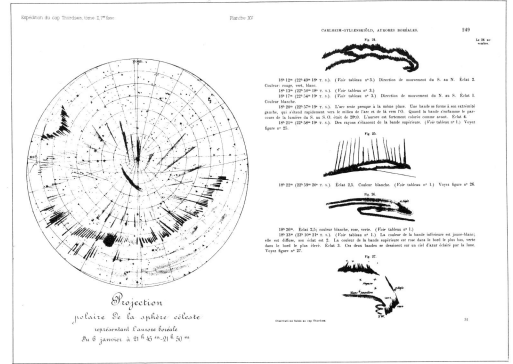

Examples of the remarkably detailed visual observations of auroras carried out by Carlheim-Gyllensköld and colleagues from Cape Thordsen, Spitzberg, during the Swedish Expedition for the First Polar Year 1882–1883.

A crayon sketch of the aurora made by
Nansen on November 28, 1893. He later used
the sketch as a basis for the facing woodcut.

Voyages of arctic exploration in the mid and late nineteenth century resulted in many beautiful engravings of the aurora.

Aurora borealis observed at Fevrior in 1874 on the Legetthof Expedition (Payer and Weyprecht).

Aurora borealis of March 21, 1879, at 2:15 A.M.—Voyage of *La Vega.*

Aurora seen by Dr. Hayes, Pt. Foulbe, on January 6, 1861.

Aurora on October 25, 1870—location uncertain.

An entirely new idea to explain the aurora was published in Austria in 1885 by Unterweger (477) and certainly must rate as one of the most hypothetical proposals of all time. Unterweger noted that a combination of the earth's velocity around the sun and the sun's velocity within the galaxy meant that there was an annual variation in the absolute velocity of the earth through space. Given the existence of a cosmic ether to explain the propagation of heat and light, this ether would be compressed at the front of the earth and rarefied behind. Then supposing that a compressed ether has a positive electrical potential with respect to the earth,

All-sky drawings of the aurora from the
Ziegler Polar Expedition 1903-1905.

Unterweger went on to deduce the direction of currents generated in the earth's atmosphere. He claimed these currents caused the aurora and that his theory accounted for seasonal and daily variation of auroral occurrence.

So although the nineteenth century saw a vast accumulation of auroral data as the result of efforts of scientists of many nations, it seemed that man was not much closer to a definitive understanding of the cause of the phenomenon. In retrospect, we see that some aspects of suggested theories were on the right track, but at the time there was no firm experimental justification for any particular idea. Nature was not to reveal the secrets of the aurora without a tougher fight!

Mistaken Aurora

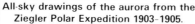

History provides numerous accounts of mid-latitude auroras (typically red and low on the horizon) being mistaken for distant fires. Seneca described an instance (414) in 37 A.D. (see page 39); a similar mistake was made in Copenhagen in 1709 when several battalions prepared to fight a fire during an auroral display (21). An Associated Press dispatch from London described a brilliant auroral display of January 25, 1938:

> The ruddy glow led many to think half the city was ablaze. The Windsor fire department was called out in the belief that Windsor Castle was afire . . . in Austria and Switzerland, firemen turned out to chase non-existent fires.

In a popular article published in 1885, Tromholt described a snakelike aurora reported by a lady in Cincinnati between 11 and 12 P.M. on March 21, 1843. The good lady claimed the snake gradually changed to a G, then an O, and then a D (529). The sketch shown here has recently been published in two books (11, 523) as an example of a curious aurora. However, a check of the *Cincinnati Chronicle* (525) shows that the event was simply a persistent meteor trail which apparently took a snakelike shape under the influence of upper atmosphere winds.

In 1941, just before Pearl Harbor, an aurora seen from Washington, D.C., was thought by some to be a new weapon being tested by the Army or searchlights helping to repel an attack by the German Air Force (236). During a World War II blackout in Los Angeles, residents saw the bright glow of an aurora above the surrounding mountains, and many called the local fire departments to report a forest fire.

The memoirs of Baron Stockmar contain an amusing story (90) of the mistake occurring in the opposite direction, with a fire being mistaken for an aurora. The anecdote concerns a Baron von Radowitz, who was apparently a verbose man, and a self-proclaimed expert on many matters of which he had little knowledge or experience. A friend of Stockmar's was traveling to an evening party near Frankfurt, where he expected to meet Herr von Radowitz. On his way he came across a barn burning and stopped his carriage to assist the people, waiting until the flames were nearly extinguished. When he arrived at the party, he found Herr von Radowitz, who had previously taken everyone to the top of the building to see an aurora, expounding on terrestrial magnetism, electricity, and related topics. Radowitz asked Stockmar's friend, "Have you seen the beautiful aurora borealis?" He replied, "Certainly, I was there myself; it will soon be over." An explanation followed as to the barn on fire. Herr von Radowitz was silent some 10 minutes, then took up his hat and quietly disappeared.

92

6

The Aurora in Colonial America

. . . the great GOD of Nature forewarns
a sinful World of approaching Calamities,
not only by Prophets, Apostles and
Teachers, but also by the Elements and
extraordinary Signs in the Heavens,
Earth and Water.

Nathaniel Ames
Article on aurora borealis in
Nathaniel Ames' Almanac, 1731

The previous chapters mention just three contributions from America to the development of auroral science, viz., the theories of Franklin and Olmsted, and the auroral zone map published by Loomis. This is about as much space as historians ever give to early American research in this field and carries with it the implication that the aurora was merely an occasional curiosity to colonial American science.

Nothing could be further from the truth. Early Americans had a great interest in natural phenomena. Though severely handicapped by the unavailability of sophisticated instruments such as those that were to be found in the established observatories of Europe, scientists in New England achieved recognition for their competence by accurate observations of the transits of Mercury and Venus across the sun's disk, by observations of the motion of comets, and for descriptions of the behavior of the aurora borealis.

But the first 100 years of settlement in New England coincided with a great sparsity of solar activity (the Maunder minimum [315]) so that auroras were seldom seen as far south as the major population centers in Massachusetts and Connecticut. It is difficult to identify the first record of the aurora during these years; *Nathaniel Ames' Almanack* of 1731 [17] claims that the first appearance of an aurora was in the year 1719 on December 11.

> Strange and wonderful have been the prodigious Effects of Nature of Late Years, in the production of terrible Thunder and Lightning, violent Storms, tremendous Earthquakes, great Eclipses of the Luminaries, notable Configurations of the Planets, and strange Phaenomena in the Heavens: the Aurora Borealis (or Northern Twilight) is very unusual, and never seen in New-England (as I can learn) 'till about 11 Years ago: Tho' undoubtedly this Phaenomenon proceeds from the concatination of Causes. For hot and moist Vapours, exhaled from the Earth, and Kindled in the Air by Agitation, according to their motion may cause strange Appearances. I do not say that this is the true Cause of these Northern Lights; but that they are caused some such way must be granted: Nor must they be disregarded or look'd upon as ominous of Neither Good nor Ill, because they are but the products of Nature; for the GOD of Nature forewarns a sinful World of approaching Calamities, not only by Prophets, Apostles and Teachers, but also by the Elements and extraordinary signs in the Heavens, Earth and Water.

An anonymous account [167] of this aurora by "an eyewitness" is prefaced by the remark, "And I hope (though I believe I will differ from some) I shall say nothing that shall be inconsistent with either Divinity or Philosophy." The description that follows refers to a "somewhat dreadful" appearance "of a flame, sometimes a blood-red color."

The most complete account of this 1719 aurora seems to be that by the Puritan minister, the Reverend Cotton Mather, entitled *A Voice from Heaven—An Account of a Late Uncommon Appearance in the Heavens* [312]. He was the greatest popularizer of science of the day, but his scientific background could not overcome his feeling that the aurora might be an omen:

> People at work about their Saw Mills, perceived their Trees to look Red with the reflection of it; and they could see to manage their work by it . . .

> It is remarkable to see how much we are left in the Dark . . . We may talk some fine things, about the Sulphur and the Nitre . . . and make ourselves be admired for our Learned Jargon, among them that have not learned the Language.

> . . . my plain Sentiments on . . . 'What Interpretation is to be made of the Aurora Borealis, that Heaven has lately shown unto us?' I will say, That though I can do very little by way of Prognostic . . . when we see a Pillar of Smoke and a Flame ascending in Heaven, we must conclude, That Evil is coming upon us . . .

Cotton Mather 1663-1728
Puritan minister of Second Church of Boston and a great popularizer of the new scientific discoveries being made in Europe. He believed in witchcraft and gathered records of it as he did of the aurora borealis, two headed snakes, and the like.

A Voice from Heaven.

AN

ACCOUNT

Of a Late

Uncommon Appearance

IN THE

HEAVENS.

With REMARKS upon it.

Written for the Satisfaction of One that was defirous to know the meaning of it.

By ONE of the Many who obferved it.

—*Rumpe Moras, Meteoraque fufpice cæli;
Illa aliquod Semper quo Monearis habent.*
Frytfchius.

BOSTON: in *N. E.*
Printed for *Samuel Kneeland,* at his Shop in King-Street. 1719.

Title page of Cotton Mather's tract concerning the 1719 aurora over New England.

Dr. Trumbull, in his Century Sermon preached in New Haven, Connecticut, on January 1, 1801, also saw fit to mention this particular aurora as signaling the first appearance of the aurora borealis in the heavens of America (296). He said that

> It filled the country with the greatest alarm imaginable. It was the general opinion, that it was the sign of the coming of the Son of Man in the heavens, and that the judgement of the great day was about to commence. According to the accounts given by the ancient people who were spectators of it, there was little sleep in New England that night.

John Winthrop 1588-1649
First Governor of Massachusetts

One might suspect a certain ecclesiastical exaggeration by Dr. Trumbull, as a review of the only newspapers printed in the Colonies at that time, *The Boston News-Letter* and the *Boston Gazette*, reveals no mention of the phenomenon at all.

In retrospect, we can attribute the claim that no auroras appeared prior to 1719 to the unusually quiet sun of the preceding 60 years. As was previously mentioned, Edmund Halley had watched patiently but in vain in England for the same reason. But clearly the aurora had appeared in New England skies for thousands of years before this date and indeed seems to have appeared at least twice during the 1600's. Two accounts from this time probably refer to the aurora: John Winthrop, the first Governor of Massachusetts, wrote in his *History of New England* (510), on January 18, 1644 (old system) (not 1643 as has been stated (204, 326)), that

Samuel Sewall 1757-1814
Chief Justice of Massachusetts

> About midnight three men, coming in a boat to Boston, saw two lights rise out of the water, near the north point of the town cove, in form like a man, and went at a small distance to the town, and to the point south, and there vanished away.
>
> The like was seen by many a week after . . . A light like the moon rose in the N.E. point in Boston, and met the former at Nottles Island, and there they closed in one, and then parted, and closed and parted divers times . . . Sometimes they shot out flames and sometimes sparkles . . .

The second account is in the diary of Chief Justice Sewall under the date December 22, 1692 (204):

> Major General [Winthrop] tells me, that last night about 7 o'clock he saw 5 or 7 Balls of Fire that mov'd and mingled each with other, so that he could not tell them; made a great light, but streamed not.

But these two auroras of the 1600's were clearly not as spectacular as that of 1719 and seemed to pass without upsetting the populace. By contrast, the 1719 aurora caused such terror throughout New England that according to the *Regents Report* (387) of the State of New York (1836) there was a suspension of "all amusements, all business, and even sleep, there being a general expectation of the approach of final judgement." The general talk of omens and impending heavenly wrath prompted Thomas Robie of Harvard to write in 1720 that

> As to prognostictions from it, I utterly abhor and detest 'em all, and look upon these to be the effect of ignorance and fancey; for I have not so learned philosophy and divinity, as to be dismayed at the signs of heaven; this would be to act the part of a heathen; not of a Christian philosopher.

Auroras remained sufficiently infrequent through the 1720's and 1730's to warrant newspaper comments when they did appear:

Edward A. Holyoke 1728-1829
Physician from Salem, Massachusetts

Isaac Greenwood 1702-1745
First Hollis Professor of Mathematics
and Natural Philosophy at Harvard
College (1727-1737)

John Winthrop 1714-1779
Second Hollis Professor (1738-1779) at
Harvard College. He was aware of a
connection between aurora
and sunspots.

On Wednesday night last we had a very bright appearance of the Aurora Borealis, or Northern Twilight and we hear that the same was so remarkable at Rhode-Island that it was surprising to the Inhabitants there.

The New England Weekly Journal
November 10, 1729

Dr. Holyoke, a physician in Salem, Massachusetts, wrote that he saw his first aurora in 1734 or 1736 when he was a young boy (296) and that "northern lights were then a novelty, and excited great wonder and terror among the vulgar." (Dr. Holyoke lived to be over 100, and in the later years of his life, from 1754 to 1828, he kept a meteorological journal in which he recorded auroras.)

The first account of aurora in New England published in a legitimate scientific journal was by Isaac Greenwood, the first Hollis professor of Mathematics and Natural Philosophy at Harvard. His 15-page article (205) in the English scientific journal *Philosophical Transactions* (in 1731) describes a brilliant aurora of October 22, 1730:

Sir, The Aurora Borealis has been very frequent with us of late; but none either for Brightness, Variety, or Duration, so considerable as what occurred on the last Thursday Night, which was the 22nd of October. This Meteor [sic] has been observed in New-England, at different Times, ever since its first Plantation; but I think at much longer Intervals than of late Years, and never to so great a Degree as the present Instance: Nor indeed is there any recorded in the *Philosophical Transactions*, that I could think, by their Description, equal to it: excepting only that celebrated one of the 6th of March, 1716 [old system], observed by the most judicious and learned Dr. Halley, and in many Respects that also must give Preference to it.

Unlike Halley, Greenwood did not attempt a physical explanation but contented himself with detailed descriptions throughout the night. Greenwood's work on aurora was short-lived, however, as in 1738 he was forced to leave Harvard, having been found guilty of "various acts of gross intemperance."

The same *Philosophical Transactions* volume published another communication (281) from the Colonies concerning this aurora (by Richard Lewis of Maryland) which ends with

Dr. Samuel Chew of Maidstone tells me, that he has for some Days past, at Morning and evening, observed several Spots, in the Sun, very plainly with the naked Eye, some of which seemed very large.

This statement implies that Lewis thought that there could be a connection between sunspots and aurora, a suggestion usually first attributed to Mairan in his book on aurora published in 1733. However, Mairan's book was written in 1731, and no doubt, many of the ideas in it were discussed before then, so it is difficult to decide who first came up with the idea. Certainly, John Winthrop, the second Hollis professor at Harvard, believed that there was some relation (260); after a description of sunspots visible with the naked eye from Boston Common on April 19, 1739, he noted that there was considerable aurora on the night of April 20. A true appreciation and acceptance of the connection of auroras and solar disturbances was not firmly established for another 100 years.

Benjamin Franklin also saw an aurora from Philadelphia in the early 1730's, and in 1737 he wrote (425) about the aurora of December 29, 1736, which appeared so red and luminous

. . . that People in the Southern Parts of the Town, imagined there was some House on Fire near the North End; and some ran to assist in extinguishing it.

Franklin was the first American to suggest an auroral theory, and his ideas were in print as early as 1749 (425). His later theory (176) (described in the preface of this book) was an extension of his earlier work. One can imagine that it must have been a restful change for Franklin to return to puzzling in the field of natural philosophy after the hectic years leading to the *Declaration of Independence.*

On April 21, 1750, a great aurora that rivaled that of 1719 was seen in America as far south as Charleston, South Carolina. An observer there wrote (195),

Benjamin Franklin 1706-1790
American scientist, philosopher, and statesman

> We had a most extraordinary appearance of the aurora borealis. One half of the sky seemed like a beautiful streaked liquid flame, so terrible to many of the female inhabitants that some of them were thrown into fits.

The first journal of science and arts to be published in the American colonies was *The American Magazine and Monthly Chronicle for the British Colonies,* founded by a "society of gentlemen" in Philadelphia in 1757. Apropos to the importance of auroral science in the Colonies, the first issue contains an anonymous article (25) on the aurora which begins

> Gentlemen
> If nothing more curious is offered you in the way of Natural Knowledge and History, for this month, than the following pieces, we hope they may receive a place in your Magazine. The first we shall offer is an attempt to solve or account for the causes of that new and beautiful phenomenon, usually called the Aurora Borealis or Northern-Light; which hitherto so much puzzl'd the conjection of modern philosophy.

The author comments on the paucity of auroras over the last century but does qualify the discussion by pointing out that they have always been common in more northern regions. When the phenomenon first appeared at the latitude of England and New England,

> ... the vulgar, ever prone to superstition, judged them a kind of prodigies and the forerunners of battles, blood, famine, plague, pestilence, and the like. Nay the sudden corruscations of their paly light, shooting and converging from the horizon to the zenith, and undulating backwards and forwards with amazing velocity, have often induced the astonished beholder to imagine that they formed themselves into all shapes, exhibiting to the eye brandished spears, and marshalled armies, and plains drenched in slaughter.

The author then demonstrates a familiarity with European theories, discussing the work of Mairan, Euler, and Halley but criticizing them for their implication that auroras should be confined to polar regions, which is not always the case. He expressed respect for the theory of Pontoppidan (page 59) as being "the most probable and ingenious" and stated that it "does not destroy the possibility of its [the aurora's] appearance in the more southerly parts." He concludes with

> But these things we submit to those who are skilled in electrical enquiries, who are more likely than any others to fix the causes of this phenomenon; and perhaps it would not be reckoned partial in us were we to flatter ourselves that, to throw new light upon this subject is an honor reserved for the curious in this country; where Electricity has already received some of its principal and most important improvements.

This anonymous work seems to be the most comprehensive foray into auroral theory in Colonial America. Accounts of isolated auroras con-

Ezra Stiles 1727–1795
President of Yale College

Jeremiah Day 1773–1867
President of Yale College

tinued to be published but added nothing to a scientific understanding. In fact statements like the following (from a "correspondent" in the first issue (118) of *American Philosophical Society Transactions,* in 1771) would seem to represent a backward step.

> ... They changed colours alternatively ... so sudden and quick that they affected the sense so strongly as to raise horror ... It was so light ... that a person, who felt no decay or infirmity of eye-sight, might easily have read a book printed in Double Pica Roman.

Benjamin Franklin's return to the contemplation of the auroral phenomenon (see the preface) must have increased European interest in the work being performed on the other side of the Atlantic. By this time, the focus of auroral research in America was centered at Yale University (322), where faculty members were keeping detailed auroral diaries. The Reverend Ezra Stiles graduated from Yale in 1746 and began to keep a meteorological journal (442) while he was pastor at Newport in 1763. He continued the journal after becoming president of Yale in 1779, and his last entry was just two days before his death in 1795. He was frustrated in his attempts to measure the height of aurora by triangulation because of the lack of a reliable second observer. On one promising occasion, his cousin John wrote that he was unable on February 30, (*sic*) 1769, to give accurate data, as he had observed the aurora "while returning on horseback from a dance, and possibly with impaired judgement."

It seems that auroras were again infrequent in New England for about 30 years around the turn of the century. President Day at Yale kept a journal (133) from 1804 to 1826, recording only 17 auroras, and commented in 1811 in his lectures on meteorology to the senior class at Yale that "so far as I have learned, scarcely any has been seen in New

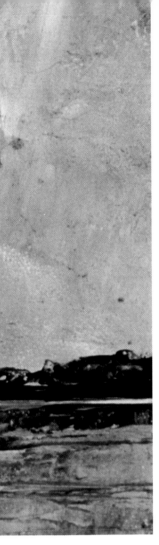

Painting by Howard Russell Butler of aurora seen in the Yukon (about 1910). This is one of five Butler paintings in the collection of the Museum of Natural History in New York.

Edward Herrick 1811–1862
Librarian and Treasurer of Yale College

Elias Loomis 1811–1889
Professor of Natural Philosophy and Astronomy at Yale College

Denison Olmsted 1791–1859
Professor of Natural Philosophy and Astronomy at Yale College

Joseph Lovering 1813–1892
Professor of Natural Philosophy at Harvard College

England for fifteen or twenty years past." The *Ladies and Gentlemen's Diary* of 1821 commented that

> The northern lights, or Aurora Borealis, for about forty years, have appeared less frequently than hereto . . . Within a few years of late, we have had few exhibitions of this sublime, and beautiful spectacle.

This return of auroras to the New England skies in the mid 1820's inspired Edward Herrick, librarian and then treasurer at Yale, to keep an auroral register (226) from 1826 (when he was only 15 years old) to 1854, as did Francis Bradley (72) from 1842 to 1854 while he was in charge of the Yale Observatory. It was primarily the journals of these men that allowed Elias Loomis (at Yale) to gather records of 4137 separate auroras and publish his map (292) of the auroral zone in 1860. The journals also permitted Loomis and Olmsted (358) (at Yale) and Joseph Lovering (296) (at Harvard) to investigate secular periodicity of the aurora. They concluded that the greatest secular periods of aurora occur every 58 (according to Loomis) to 65 years (according to Olmsted and Lovering) with a duration of maximum of 20–25 years.

Thus auroral studies constituted an important part of colonial and post-Revolutionary science in America. But with the infrequency of the phenomena at the latitudes of northern American states, little progress could be made from the mid-nineteenth century state of understanding unless expeditions were organized to the arctic regions. Such expeditions would have been costly and time consuming, and active auroral research diminished in the late part of the century. The next advances in the field were to come from countries located nearer the auroral zone, and it was Norwegian scientists who pioneered the twentieth century advances in our understanding of the aurora.

While working at Harvard College Observatory in the 1860's
and 1870's, Trouvelot made many original drawings of
interesting celestial objects and phenomena. In 1876 these
drawings were displayed at the U.S. Centennial Exhibition at
Philadelphia, forming part of the Massachusetts exhibit. Some
of the drawings from the exhibition were published in 1882,
and included in the collection is this drawing
of the aurora borealis.

"Aurora Borealis" by Frederick Church (1865).

Another of Church's arctic paintings, "Icebergs," recently (October 1979) sold for $2,500,000, the highest price ever paid for an American painting.

Frederick Church 1826–1900
American painter

7
Legends and Folklore

He knew, by the streamers that shot so bright,
That spirits were riding the northern light.

Sir Walter Scott
The Lay of the Last Minstrel, 1802

One wonders why history does not tell
us of 'aurora worshipers,' so easily
could the phenomenon be considered the
manifestation of 'god' or 'demon.'

Robert Falcon Scott
Scott's Last Expedition, 1913

Consultation of standard reference works and dictionaries of folklore and mythology gives the impression that little has been written on the influence of the aurora borealis on the beliefs of northern people. It would seem surprising that the overwhelming presence of such a striking display in the night skies should go neglected by storytellers, and indeed, a patient search of the literature is rewarded by reports of a variety of traditions concerning the aurora. It does seem though that little effort has been made to collect systematically and document adequately these traditions and that the topic might provide a fruitful subject for anthropological and ethnological research.

There are many practical interpretations of cosmogenic weather predictions (352), especially in Norway and Finland, that relate auroras to both good and bad weather, to the coming of winds, to changes of weather, and even to longtime weather forecasts (see page 125). That great auroras foretell the death of kings was a common belief in Scandinavia in the fifteenth to eighteenth centuries; in Germany in the same period the aurora was viewed as a transcendental happening to warn people to repent and to mend their ways (62).

Perhaps the most common belief regarding prognostication by auroras is that great auroras foretell war; this tradition may be found in folklore from Poland, Prussia, Germany, Denmark, England, Estonia, and Lapland (209). The Lapps believed that auroras occurring all over the sky foretold terrible war (301). After the French-German war, auroras seen over Malta were said to be reflections from French blood in the sea.

In the realistic tradition, we find interpretations in terms of reflections of action taking place in far-off locations, e.g., in Norway, reflections of schools of silver herring in distant oceans, or icebergs in the North Atlantic, or geysers in Iceland and in Estonia, reflections from a dreadful creature chained on a far-off moor (352). In parts of Lapland the northern lights were simply seen as a winter analogue of summer's thunderstorms, both phenomena being assumed to be a mixture of fire and water (301). A

An oil painting by Carl Bock made by the light of the aurora at Porsanger Fjord, Norway, on October 3, 1877.

(Opposite page), Finnish legend says that when *Repu* the fox strikes the snow with his tail, he creates a fire of no heat which is the aurora borealis.

Finnish story says that the aurora is caused by a big fish or whale romping in the Arctic Ocean; when the whale blows water, sunbeams stick to it and cause the Arctic to burn, and when the whale strikes the water with its tail, great waves of auroral light splash out. An Estonian myth also relates the aurora to a whale playing, its scales flashing in the sun, and to crocodiles playing in the Arctic Ocean. Martti Haavio, a Finnish ethnologist, believes that the whale and crocodile legends were inspired by Behemoth or Leviathan (*Old Testament: Job 40, 41*), as is probably the legend from Finland concerning Peemut (209):

> God created an animal Peemut, who cannot move, and lives in a swamp covered in moss and brush. Small animals run into its mouth, and ravens also fly into it. When the animal moves or wheezes, it causes aurora. When it knows bad weather is coming, it makes auroras. If it could get free it could make people boil like water in a kettle, and if it blew on black coal it would turn red and burn.

The modern Finnish word for aurora is *revontuli,* which has its origin in another legend that attributed the aurora to the fox, *Repu.* Revontuli means "fox-fire," and the legend recounts that when Repu turns, his fur flashes; he strikes fire with his tail but a fire of no heat. Folklore says that Repu's fur can be used for light in a gunpowder cellar (209).

The Miracle of Our Lady of Fatima

One of the most interesting examples of the aurora as a prognostication occurred in modern times. On May 13, 1917, in the little hamlet of Adjustrel in Fatima, Portugal, three shepherd children claimed to have seen a figure of a lady, brighter than the sun, standing on a cloud in an evergreen tree. They were instructed by her to return on the thirteenth day of every month, and by October 13, they were accompanied by a crowd of 50,000. Only the children saw the lady each time, but on October 13, others claimed that the sun rotated violently and fell and danced over their heads before returning to normal. On this occasion the lady revealed herself to the children as Our Lady of the Rosary, Mother of God, and offered the following prophecy (320):

> This war [World War] is going to end, but if people do not cease offending my Divine Son, already too grievously offended, not much time will elapse and during the Pontificate of Pope Pius XI, another and more terrible war will begin. When you will see a night illuminated by an unknown light, know that this is the great sign that God gives you and that the chastisement of the world is at hand.

The incident was investigated in great detail by the Catholic Church and is referred to as the Miracle of Our Lady of Fatima. And it was during the Pontificate of Pius XI that the other more terrible war, World War II, began. And seen all over Europe on January 25, 1938, from 9 o'clock in the evening until 2 the next morning, the heavens were "filled with a strange and terrible crimson fire." It was, of course, a display of the aurora borealis of unusual splendor, but many Catholic theologians objected to science explaining away the "unknown light" and drew the prophetic connection when, 3 months later, Hitler invaded Austria and set the stage for the great chastisement predicted at Fatima.

The more spiritual traditions found in the arctic regions are most interesting in that they reveal an astonishing homogeneity in their main traits. This similarity was pointed out by the Norwegian ethnologist Nordland and cited as evidence supporting possible migration paths of circumpolar cultures (352). Invariably these traditions involve death or the spirits of people who have died. But not all spirits are allowed to flicker in the heavens, and it is the requirements for admission that vary with different arctic cultures.

The Eskimos of the Hudson Bay area of Canada tell this story (352):

The ends of the land and sea are bounded by an immense abyss, over which a narrow and dangerous pathway leads to the heavenly regions. The sky is a great dome of hard material arched over the earth. There is a hole in it, through which the spirits pass to the tree heavens. Only the spirits of those who have died voluntary or violent death, and the raven, have been over this pathway. The spirits who live there, light torches to guide the feet of the new arrivals. This is the light of the Aurora. They can be seen there, feasting and playing football with a walrus skull.

(This Eskimo folklore that associates the aurora with the spirits or souls of the deceased at play, using a walrus head for a football, was reported as early as 1745 by the Danish missionary Hans Egede (152) in his *Descriptions of Greenland.*) The Eskimos of Nunivak Island had the opposite idea of walrus spirits playing ball with a human skull (385).

The Norwegian explorer Nansen tells us (344) that the Eskimos of the west coast of Greenland believed that when the soul left the body upon death, it went either to a place under the earth and the sea or to the up-

107

Auroral band over an Eskimo igloo.

Eskimo soapstone carving—(Top) "Northern Lights in Povungnituk" by David Davidialuk (Povungnituk, Quebec, Canada). The head represents the spirit of the northern lights in the serpentlike auroral form. (Bottom, left) Face watching the northern lights (carver unknown). (Repulse Bay, Northwest Territory, Canada). (Bottom, right) Two old men watching an extraordinary display of the northern lights. Carver unknown. All carvings are in the Eskimo Museum, Churchill, Manitoba, Canada.

Davidialuk (died 1976) was known as a storyteller 'par excellence.' His carvings were considered crude until he was recognized by outside specialists.

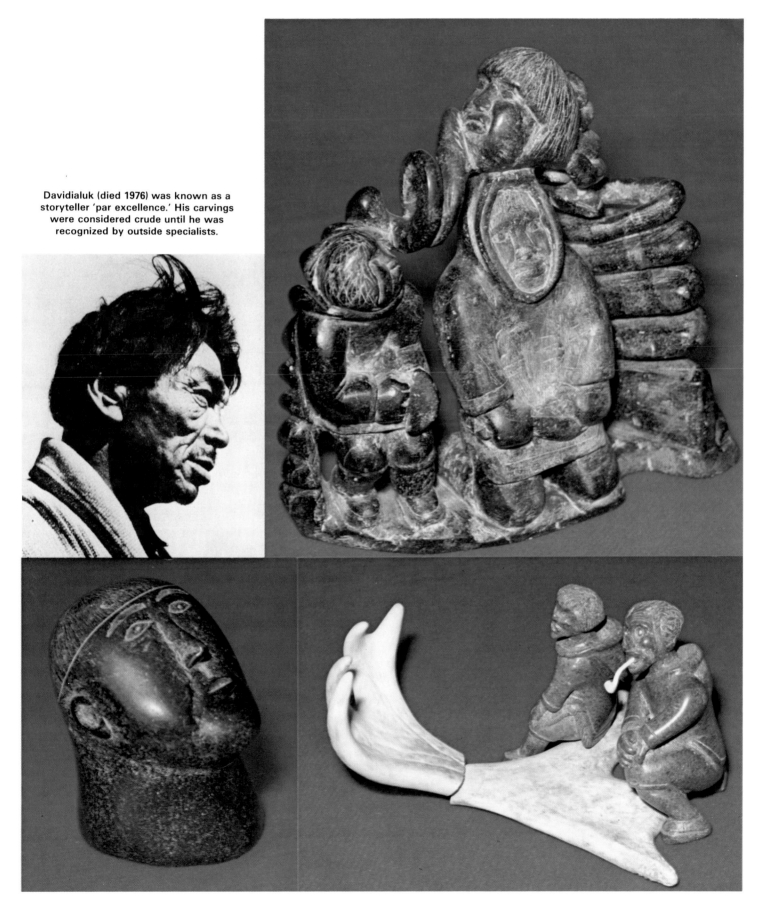

per world in the sky. The Eskimos have no hell, and both these places are more or less good. But the

> ... overworld is colder ... over it is arched the blue heaven. There the souls of the dead dwell in tents around a lake, and when the lake overflows it rains on earth. There are many crowberries there, and many ravens who always settle on the heads of old women and cling onto their hair ... the souls of the dead can be seen up there by night, playing football with a walrus head. Those who die ordinary deaths descend to the world below, where they carry on a monotonous existence, which is free, however, from the cold and hardship of their earthly home.

Rasmussen told of a similar belief in another Eskimo culture in Greenland (384):

> It is this ball game of the departed souls that appears as the aurora borealis, and is heard as a whistling, rustling, crackling sound. The noise is made by the souls as they run across the frost-hardened snow of the heavens.

This interpretation of the aurora as the spirits playing football is widespread among the Eskimo tribes (384); the Eskimo word for aurora, aksarnirq, translates into "ball player." Naive simplicity characterizes many explanations Eskimos propound for complex natural phenomena. They rarely need to pursue an enquiry more than one step; the Eskimo accepts the all-sufficient instrumentality of spirit agencies and so requires no other explanation.

Eskimos of the extreme north of Canada speak of spirits of the dead clothed in an ethereal light, amusing themselves in the absence of the sun; they call bright everchanging aurora the "dance of the dead." In east Greenland the stories are more macabre; there the aurora is seen as the spirits of stillborn children or premature children who were put to death (494). Then children's souls

> ... take each other's hands and dance around in mazy circles. They play at ball with their afterbirth, and when they see orphan children, they rush upon them and throw them to the ground. They accompany their sports with a hissing, whistling sound.

Thus the aurora is also called alugsukat, which appears to mean untimely births or children born in concealment. This notion of the Greenlanders seems to be related closely to the Indian belief that the northern lights are the dead in dancing array.

Some Eskimo tribes claim that they hear the aurora (see page 158) and imitate the sound by whistling in such a manner that the aurora is attracted and comes closer, so that they can whisper messages to the dead (385). The folklore that says the aurora will come closer if one whistles is widespread throughout the Eskimo and Indian regions of North America. Mr. Mercredi, an Indian of Canada's Northwest Territories, tells the story (207):

> When I was a boy, people respected the northern lights as messengers of our spirits. People also said that some men and women can bring the northern lights down and make them obey their commands. You only have to walk outside, rub your hands together and whistle, then the northern lights will come down and sing and dance for you. When I was a young man I tried it once and it worked.

The aurora often appears flashing and flame like, and some native beliefs concern heavenly fires. The Makah Indians of Washington thought they were fires in the north lit by a tribe of dwarfs who were so strong that they caught whales with their bare hands (455). The Mandan of North Dakota saw the aurora as fires over which the great medicine men and warriors of northern lands simmered their dead enemies in enormous pots. The Menomini Indians of Wisconsin regarded the lights as torches used by great friendly giants in the north to spear fish at night (385). The Tlingit Indians of the northern Pacific coastal area, the Chuckchee in Siberia, and the Lapps of the USSR and Finland all speak of the souls of those who have died violent deaths continuing to fight with one another in the air (303). The frequent red lower borders of the aurora are related to bloodshed in these conflicts (352).

Other Indian tribes explain the aurora through lengthy myths. The lights were thought to represent some action of legendary heroes. The aurora plays the role of advisor to Chief He-Holds-the-Earth in Onondaga Indian genesis myth (303). The Ottawa Indians see the aurora as reflection of a great fire occasionally kindled by their creator Nanahboozho, reminding them of his interest in their welfare (212).

The Dog-Rib tribe of Chippwyan Indians recounts the legend of Ithenhiela (the Caribou-footed), an escaping servant boy who threw down a clod of earth that became hills, a piece of moss that became a muskeg swamp, and a stone that became the Rocky Mountains. Finally, he climbs a tree into the Sky Country and assists recovery of the medicine belt that allows the sun to shine. As a reward, the great chief gives his daughter, Estanda, and they live forever in the sky. When the Northern lights dance across the sky, the Indians see in them the fingers of Ithenhiela beckoning them to the home he found so far away (46).

A legend of the Pacific Northwest Indians (411) tells of a chief's son who could travel to the Land of the Northern Lights, where he played ball with the Northern Light people. Once the chief followed his son there and watched, and they returned to the Lower World on two great birds flying down the Milky Way. However, all memory was erased from the chief, and except for the Chief's son, those very few who travel to the Land of the Northern Lights do not remember their remarkable journey.

Some North American Indian tribes call the aurora "the old man," who is described as having long white hair, eyes of fiery brightness, and a robe of skins around his shoulders. The streamers of auroral light suggest the idea of the old man's hair, the flashing of aurora corresponds with brightening of his eyes, and the robe of skins is represented by the auroral glare.

The pioneer of arctic Canada, Samuel Hearne, wrote (218) that the Indians there "say their deceased friends are very merry" when the aurora is bright and are "dancing in the clouds." This conception was used by Longfellow (290) when he had the boy Hiawatha see

. . . The Death Dance of the Spirits,
Warriors with their plumes and war-clubs
Flaring away to the northward
In the frosty nights of winter.

Reports that the aurora was physically feared by the Eskimo cultures are not common. Eskimos near Point Barrow, Alaska, however, always carried a knife for protection and threw dogs' excrement and urine at the lights (494).

111

Facsimile of an old drawing depicting Lapps
hunting by the light of the aurora borealis.

A tradition among Norwegian Lapps relates that 'when two Lapps begin to quarrel, they would sit down on the ground and sing to be sent light, and when the lights met in the sky, they fought with one another, accompanied by a terrible noise and crackling. How this light acted against unpleasant behavior is told in the following story (352): After two brothers had kindled fire and eaten,

> . . . a bright northern light appeared. Then the youngest brother started chanting
>   'The northern light runs . . . lip, lip, lip
>   with fat in the mouth . . . lip, lip, lip
>   with a hammer in the head . . . lip, lip, lip
>   with an ax on the back . . . lip, lip, lip'

His brother forbade this, but he started singing even louder. Then the northern light started to flicker so strongly that it cracked against the snow like a hard skin cracks. The elder brother crept under his wooden sledge, but the aurora borealis killed his younger brother and burned his fur coat . . . even the body of the boy disappeared.

Lapps in Finland recite a similar poem

> Lappland jays running . . . lip, lip, lip
> Axe on their heads . . . lip, lip, lip

and liken the aurora to the Lapland jay flying across the sky with an axe. The aurora is said to have killed people with this axe. Old Lapps to this day believe that the aurora borealis is not to be mocked, or it will come down and burn them. For this reason they will not use sleigh bells under an aurora nor make unnecessary noise in their *Lappeot* (dwelling) when the aurora is bright. Women will not go bare headed if the aurora is bright, as tradition has it that the aurora can seize onto their hair (301).

The Russian Lapps declare the aurora to be the spirits of the murdered. These spirits live in a house, where at times they gather and begin stabbing one another to death, covering the floor with blood. They are afraid of the sun and hide themselves from its rays. The aurora appears when the souls of the murdered begin their slaughter and was feared by the Lapps (303).

112

According to old Lapp tradition, it is dangerous to travel under the aurora as depicted here, with a bell around the reindeer's neck.

Tradition that the Lapps regularly use the light of the aurora to travel by is illustrated in these two drawings.

The Laplanders still believe it is dangerous to whistle at the northern lights or to *jojka* (a Lappish way of singing similar to yodeling) at them. Extended gazing at the aurora is also warned against as it is told that the lights will then get brighter and come closer and possibly tear out one's eyes (301).

The American Fox Indians also feared the aurora somewhat, believing it to be flames flashing upward from the place where the northern sky meets the earth. They are the ghosts of slain enemy trying to rise, restless for revenge; the sight of them was regarded as an ill omen, a sign of fighting and pestilence (303).

Direct reverence to the aurora as a god seems only to be found in the beliefs of the Chuvash, a Siberian Indian tribe. Their heaven god, *Suratan-Tura,* was also their name for the aurora, and he was associated with birth and helping women through the agonies of childbirth. They believed that the sky gave birth to a son during a bright aurora (303).

But by far the most widespread interpretation of the aurora among the more primitive northern people is that it is the spirits of the dead either fighting or playing. This special after-life existence seems reserved for those who have died violent deaths and thus represents an extension of one of the general functions of religions by serving as a comfort system for those members of the tribes who take special risks. Their final abode is to be found upward and northward, where they continue their games and battles in the aurora borealis.

The more developed mythologies of advanced cultures in or near the arctic also have interpretations of the aurora. In Scotland, there is a tale (305) of the *slaugh,* or spirit world, which is constantly at war and may be seen and heard on clear frosty nights in the *fir chlis* (aurora borealis). Stories speak of the "merry dancers" engaged in clan fights for a fair lady.

Scandinavian mythology draws an image of the aurora as the beard of the war god Thor. Longfellow (291) takes up this theme:

> The light thou beholdest
> Streams through the heavens
> In flashes of crimson,
> Is but my red beard
> Blown by the night wind
> Affrightening the nations.

> Henry Wadsworth Longfellow
> *The Saga of King Olaf, 1:*
> *The Challenge of Thor*

The rich Norse mythology associates the aurora with the most beautiful of all goddesses, Freya, goddess of beauty and love (340). But Freya was not soft and pleasure loving alone, for she often led the Valkyries down to the battlefield. The name Valkyrie means "choosers of the slain," and Odin, king of the Norse gods, mounted these warlike virgins on horses and armed them with helmets and spears, sending them down to the battlefield to choose who would be slain (275).

> When they rode forth on their errand, their armour shed a strange flickering light which flashed over the northern skies, making what we mortals call the aurora borealis or northern lights.

The Latvians also related the aurora to the spirits of fallen soldiers (*Murgi*) and souls of the dead (*Iohdi*) battling in the air (303).

The appearance of aurora along the arctic shores of Siberia through the centuries has resulted in a wealth of folklore from these areas. The Yukagirs and Yakuts, people of this region, tell stories (454) that attempt to explain the aurora:

> In ancient times, Yukagirs living on the Yana River worshipped a silver scaly fish which came yearly from the sea to the mouth of the river, and the people made rich sacrifices to the fish. Once there was a year of scarcity and Yukagirs had nothing to sacrifice to the fish. When the fish approached the shore, no one came to meet it. The fish became angry and with a mighty blow of its tail spilled the waters into the sky where reflections from it began to sparkle in the heavenly dome to the north. From that time the Yukagirs believe that when the sky is sparkling and flaming [with aurora], bad weather is in store for them.

The legend of the Yakuts people states that

> Once upon a time the Yakuts attacked the Yukagirs and drove them away to sea. The pursued Yukagirs burnt a great number of fires which flamed so high that it seemed the whole heavens were burning.

Consequently, the Yakuts call aurora Yukagirian fires.

Russian Lapps have given a completely mythological explanation to the reddish tint seen in the northern lights. A mysterious male figure by the name of *Naainas* is killed when the sun's rays hit him by mistake, and the auroral light arises from his blood (301).

It is unfortunate that the auroral folk culture of the northern peoples is scattered through many old and obscure literature references. Certainly, these stories have essentially disappeared in the modern world, with only the very old people recalling what must have been a rich verbal tradition. The incredible religious traditions developed in the north involved prophets, sacral kingship, fertility gods, ancestor worship, communication with the spirits, and widespread shamanism, where the traveling shaman was the mystic intermediary between man and divine power. The characteristic flickering lights of the polar skies were assimilated into all these aspects of religious belief and formed an essential part of arctic culture up until the present century.

The Hollow Earth Theories

Halley and Euler seem to have been the latest scientists of note to suggest that the earth might contain an inner concentric (and perhaps habitable) sphere. However, ideas of an inner earth have resurfaced many times since the 1700's. The June 18, 1818, issue of the *Daily National Intelligencer* (Washington, D. C.) contained a declaration by John Cleves Symmes (531), a former captain of the U.S. infantry, that

> the earth is hollow, habitable within; containing a number of concentrick spheres; one within the other, and that it is open at the poles twelve or sixteen degrees

The pronouncement was accompanied by a certificate of Symmes' sanity and his pledge to devote his life to commanding an exploring expedition to prove he could enter the inner earth. He diligently studied explorers' journals and scientific papers for evidence to prove his theory and believed that the aurora resulted naturally from his idea. Symmes thought that the aurora borealis was produced by rays of the sun projecting into the inner world through the southern opening; by two refractions at each opening and by reflections from the inner concave surface, these rays were seen in the north as rays of the aurora.

The theory of a hollow earth with polar openings was revived again in 1906 by William Reed in his book *Phantom of the Poles* (532). He propounded that the aurora was the reflection of fires from the interior of the earth. The exploding and igniting of a burning volcano containing minerals and oils would cause much of the coloring, whereas an absence of coloring or a faint toning would be caused by burning vegetable matter such as prairie grass or forests.

In 1920, Marshal B. Gardner published *A Journey to the Earth's Interior, or Have the Poles Really Been Discovered?* (526). He differed from Symmes and Reed in that he believed there was a central sun within the earth which was (among other things) the source of the aurora borealis. The rays of this central sun would project through the polar openings, and the changing forms and streams of the aurora would then result from passing clouds cutting off the rays.

More recently, the theory of a hollow earth has been seriously propounded (522) by Raymond Bernard in his book *The Hollow Earth* written in 1963. He maintains that the polar openings into the inner world were discovered by Admiral Byrd on his polar flights and that there is a high-level Defense Department conspiracy to keep this knowledge secret. Bernard strongly supports Gardner's interpretation of the aurora, and states that it is not a magnetic or an electrical disturbance but simply a dazzling reflection from the rays of the central sun. (Bernard has written many books on a wide range of topics: *The Unknown Life of Christ, Flying Saucers from the Earth's Interior, Herbal Elixers of Life,* and *Constipation.* His books are only available in mimeograph or xerographics.)

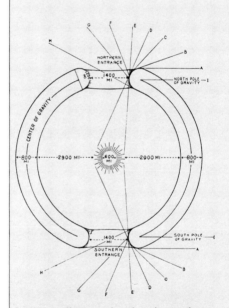

Marshall B. Gardner 1854–1920
American author of hollow earth theory

Gardner's diagram showing the earth as a hollow sphere with its polar opening and central sun. A traveler on a journey to the earth's interior would see the aurora from point A onward. At point D he would catch his first glimpse of the corona of the central sun.

8
Aurora Bright, Rain Tonight?

And now men see not the bright light which is
in the clouds:
But the wind passeth and cleanses them.
Fair weather cometh out of the north:
With God is terrible majesty.

Book of Job, 37:21–22, ~425 B.C.
(Original King James Translation)

And now men cannot look at the light
when it is bright in the skies
when the wind has passed and cleansed them
Out of the north comes golden splendor,
God is clothed with terrible majesty.

Book of Job, 37:21–22, ~425 B.C.
(Revised Version)

Although the mystery and beauty of the aurora has fascinated scientific and nonscientific observers alike throughout the centuries, the rare occurrences of auroras at the mid-latitudes of population concentrations gave little opportunity for men of science to study the phenomena. As a result, theories and hypotheses on the nature of the aurora were often based on marginal observational evidence, so that the auroral literature came to contain a plethora of misconceptions. It is the nature of man when confronted with something he cannot understand, to try to relate it to more familiar concepts. In the nineteenth century, many attempts were made to connect the aurora with various meteorological and astronomical phenomena.

The shape of auroral forms is often similar to formations seen in cirrus clouds so it is not surprising that Dr. Richardson on Franklin's arctic expeditions concluded (177) that "the aurora borealis is constantly accompanied by or immediately precedes the formation of one or other of the various kinds of cirrostratus."

The *Southern Cross* expedition which wintered at Cape Adare (72°S) in 1899-1900 found that brilliant and active aurora observed to the north was followed "too often for coincidence" by a violent storm from the southeast (51).

The ninth edition of the *Encyclopedia Brittanica* reports an observation of transition back and forth between cirrus clouds and aurora, the one transforming to and then reforming from the other.

Many observers found that the alignment of cirrus clouds was perpendicular to the magnetic meridian and thus concluded that the cause of this orientation also directed the alignment of the aurora. Fritz and Weyprecht independently stated that every time polar bands of clouds were observed during the day there was an aurora the following night (21). Fritz in particular was convinced of a connection between the aurora and the weather (179):

Representation of auroral radiation and its association with cirrus clouds.

. . . while the isochasm [lines of equal probability of auroral occurrence] did not exhibit as close an agreement with the isoclimics [lines of equal inclination of the magnetic field lines] there did seem to be a relation to the isobaric curves, the region of lowest pressure being the regions of maximum auroral frequency.

Writing in the *Philosophical Transactions,* a Mr. Winn claimed (509) that in 23 instances, without fail a strong gale from the south or southwest followed the appearance of an aurora. If the aurora was bright, the gale came on within 24 hours but was not of long continuance; if the light was faint and dull, the gale was less violent and longer in coming but also of longer duration. His observations were made in the English Channel, where such winds are very dangerous, and by attending to the aurora, he says, he often escaped shipwrecks, while others suffered.

These are but a few of the nineteenth century observations concerning the aurora and clouds that led Angot in his book *The Aurora Borealis* (21) (published in 1896) to the conclusion that

Alfred Angot 1848-1924
French scientist, Director Central Bureau for meteorology, Paris, and author of *The Aurora Borealis* published in 1896

> . . . an undoubted relation exists between certain clouds of the cirrus class and the aurora borealis; the two phenomena are subject to the same laws of periodicity, succeed each other or even coexist, their analogy being often so close that many observers are of opinion that the appearance of the aurora depends on the presence of these clouds. We shall see that these facts are of great importance to the theory of the aurora.

The *Encyclopedia Brittanica* at this time asserted that auroras were produced by the same general phenomena as are thunderstorms and concluded that the auroras of 1859 and 1869 assumed the character of thunderstorms which

> . . . instead of bursting into thunder, had been drawn into the upper parts of the atmosphere, and their vapor being crystallized into tiny prisms by the intense cold, the electricity became luminous in flowing over these icy particles.

Piazzi Smyth 1845-1888
Famous Egyptologist and Astronomer Royal for Scotland

However, Professor Smyth at the Royal Observatory, Edinburgh, then published observations (435) showing that the monthly frequency of aurora varied inversely with thunderstorms.

It is not unusual to find similar instances of completely contrasting opinions in the literature on the association of the aurora with the weather. In Labrador, for instance, it was said that colored auroras foretold fine weather, whereas in Greenland they tended to herald the south wind and storms. In Christiania, Norway, Hansteen concluded from a long series of observations that the aurora was nearly always followed by a lowering of the temperature, yet nearby at Abo and Helsingfors, Argelander maintained that the temperature rose during auroral activity (213).

Christopher Hansteen 1784-1873
Norwegian astronomer and mathematician

Many reported that the barometer tended to fall after an auroral display, and thus the aurora foretold bad weather. The Royal Society of Edinburgh in 1868 mentioned the fact as important to agriculturists, stating that an aurora following a period of fine weather is "a sign of a great storm of rain and wind in the afternoon of the second day afterwards." But if a farmer then read Dalton's book *Meteorological Observations and Essays* (128), he would have been confused, as Dalton concluded from a statistical study that the appearance of the aurora was a prognostication of fair weather.

Dalton was better schooled in mathematics than many others who concerned themselves with aurora and weather, and his reasoning to the

John Dalton 1766-1844
English chemist and natural philosopher

above conclusion makes an interesting argument:

Since the Spring of 1787, there have been 227 aurorae observed at Kendal and Keswick; 88 of the next succeeding days were wet, and 139 fair, at Kendal; now, in the account of rain, the mean yearly number of wet days there is stated at 217, and of course the fair days are 148; hence the chances of any one day, taken at random, being wet or fair, are as these numbers. But it appears the proportion of fair days to wet ones succeeding the aurorae, is much greater than this general ratio of fair days to wet ones; the inference therefore is, that the appearance of the aurora borealis is a prognostication of fair weather.

The only objection to this inference which occurs to us as worth notice is, that the aurora being from its nature only visible in a clear atmosphere, this circumstance of itself is sufficient to cast the scale in favour of the succeeding day, being fair, without considering the aurora as having any influence either directly or indirectly. — The objection has undoubtedly some weight; but upon examining the observations, it appears that the aurora not only favours the next day, but indicates that a series of days to the number of 10 to 12, are more likely to be all fair, than they would be without this circumstance.

Of 227 observations, 139 were followed by 1 or more fair days, 100 by 2 or more, etc. as under.

1	2	3	4	5	6	7	8	9	10	11	12
139	100	69	52	38	30	21	16	10	6	2	1

According to the laws of chance, the probability of any number of successive fair days is found by raising 148/365 to the power, whose index is the proposed number of fair days; these probabilities being multiplied by 227 will give what the above series ought to have been, if the aurora had no influence; it is as under.

1	2	3	4	5	6
92	38	15	6	2	1

From which it appears there should not have been more than 1 aurora out of 227 followed by 6 fair days, and yet in fact there were 30; whence the inference above made is confirmed.

Engraved from a drawing made on the French Arctic Expedition to Bossekop, Finland, 1838-1840. Seen to the south, January 18, 1839, at 7:40 P.M.

(Opposite page) Engraved from drawings made on the French Arctic Expedition to Bossekop, Finland. 1838-1840.
Seen to the north, January 6, 1839, at 6:27 P.M.
Seen to the south, January 6, 1839, at 6:04 P.M.

**Alexander McAdie 1863–1943
U.S. Signal Service meteorologist and
later Director of Harvard's
Blue Hill Observatory**

The most ambitious exposition of the relation of the aurora to meteorology was commissioned by the U.S. Signal Office of the War Department and published by Alexander McAdie (318) in 1885. The article presents "emphatic proof to support the statement that only when an area of low pressure is closely pressed by an area of marked high is an [auroral] display likely to occur."

McAdie claimed to confirm Fritz's contention that low-pressure areas cover the places of maximum auroral frequency. Though he was aware that such a relation did not hold in equatorial regions, McAdie still regarded this as a significant cause-effect relationship. An analysis of thunderstorm activity led him to the conclusion that "both were different manifestations of the same action, due to similar causes and varying in appearance because of the different [meteorological] conditions at the places of occurrence." The optimum meteorological conditions to promote auroras were described as not too dry and not too moist; arid areas were free of both thunderstorms and aurora, and very moist climates would not allow the accumulation of electrical charge.

McAdie's most quantitative conclusion concerned the brilliancy and color of auroras, which he said directly depended on the steepness of the barometric gradient. Small gradients were supposed to favor pale bluish or yellowish auroral light, while deeper red colors and more brilliant streamers indicated a more stubstantial pressure gradient. Finally, McAdie was convinced that feathery cirrus clouds invariably formed as a result of an auroral display.

In retrospect, it would seem well that the Chief Signal Officer took the prudent step of placing a note at the beginning of McAdie's article stating that "the Chief Signal Officer does not thereby necessarily endorse the views set forth."

These nineteenth century ideas on relations between aurora and meteorological phenomena have been viewed with amusement and condescension by later scientists, and indeed once the height of the aurora had been established beyond doubt as exceeding 60–70 miles, there seemed little possibility of a connection with weather phenomena contained below 10 miles.

Recent years have seen an increasing interest in the possible effects of magnetic storms (implying associated aurora) and the weather. Walter Orr Roberts, then of the National Center for Atmospheric Research in Colorado, stated (392) in 1976 that:

> We find some fairly convincing evidence (although not all of our colleagues are yet convinced) that if there is a large magnetic storm there are worldwide effects at the jet stream level of the atmosphere, with low pressure centers that correspond to major storms over the Gulf of Alaska, deepening in the first two to four days . . . There seems no doubting that there is a weather effect . . . over large parts of the world.

The past 2 years, 1977 and 1978, have seen sessions at major geophysics conferences devoted to possible relations between weather, climate, and solar activity, as well as an entire 1-week conference devoted to just this topic (see page 124). But certainly any such connections will be found to be on a worldwide scale. It is very improbable that local auroras could ever relate to local weather conditions, and the many claims of this over the past 200 years must be attributed to "old wives' tales," on the part of the general public, and to poor statistical analyses on the part of scientists.

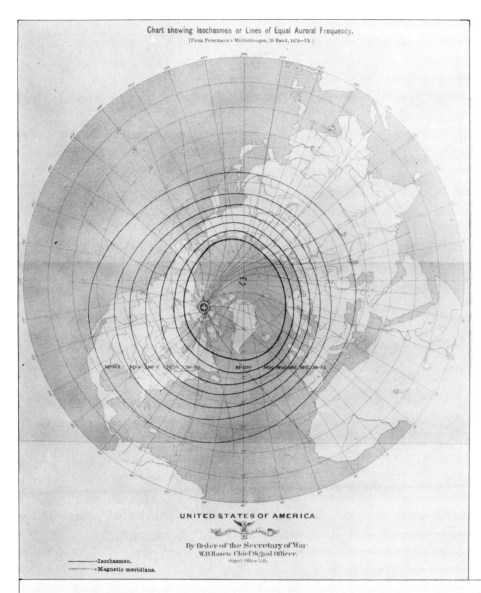

Chart showing Isochasmen or Lines of Equal Auroral Frequency.
[From Petermann's Mittheilungen, 20 Band, 1874—IX.]

UNITED STATES OF AMERICA.

By Order of the Secretary of War.
W.B.Hazen, Chief Signal Officer.
Signal Office Lith.

——— =Isochasmen.
········· =Magnetic meridians.

Early U.S. military interest in the aurora concerned the possible relations between the aurora and weather. This diagram was commissioned by the Secretary of War.

Weather map purporting to demonstrate that auroras occur were there is a sharp gradient in the barometric pressure. Map prepared by U.S. Signal Corps. The aurora on this date was visible from New England, Minnesota, and Michigan on September 3, 1883.

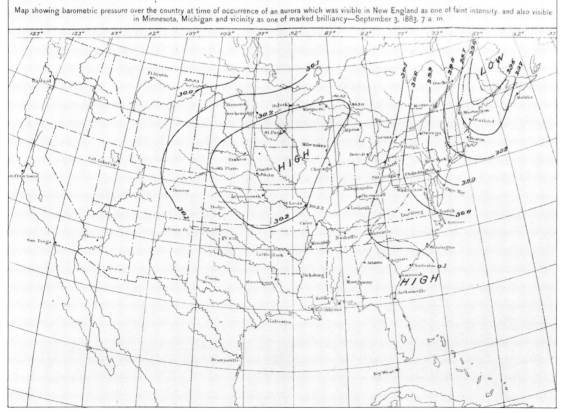

Map showing barometric pressure over the country at time of occurrence of an aurora which was visible in New England as one of faint intensity, and also visible in Minnesota, Michigan and vicinity as one of marked brilliancy—September 3, 1883, 7 a. m.

SOLAR-TERRESTRIAL INFLUENCES ON WEATHER AND CLIMATE

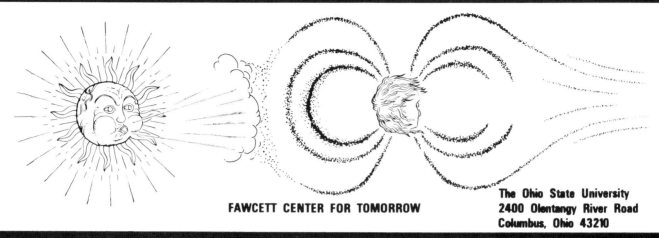

FAWCETT CENTER FOR TOMORROW

The Ohio State University
2400 Olentangy River Road
Columbus, Ohio 43210

JULY 24-28, 1978

Increasing interest in possible associations between the sun, aurora, and the weather is evidenced by this recent announcement of a 1-week conference devoted to the subject.

The eighteenth and the nineteenth centuries also saw claims of associations between auroras and other astronomical bodies; an essay by Dalton (128) in 1789 claimed that the aurora occurred more frequently around new moon than at any other time, which suggested that atmospheric tides occasioned by the moon might have some influence.

Over 100 years later, Arrhenius also claimed the moon affected auroral occurrence and suggested an ingenious explanation. He said the moon is not protected by an atmosphere from the stream of negative particles from the sun, so will reach a very high negative potential. Thus the moon would seriously affect the number of auroras on any place over which it is located, as it would lower the potential gradient in the earth's atmosphere and reduce the number of discharges that caused aurora (31).

Frequent remarks (90) claiming that the brightness of Venus and of some stars appeared greater when viewed through an aurora, "acting as a screen and removing the glare with which so bright an object as Venus is always accompanied." A number of independent observers claimed that Jupiter's belts exhibited the brightest colors during periods of aurora, and during the period 1876-1878 when few aurora were seen, Jupiter's equatorial zone and belts were reported to have mainly light tints (90).

Various spectroscopic observations of the aurora and the zodiacal light in the 1860's and 1870's suggested similar spectra, and Angström and others theorized on the identity of origin of the two phenomena. Such thoughts were probably inspired by Mairan's original writings on the subject. Better observations with improved instrumentation beneath the clear skies of Italy led professor Smyth (434) and others to refute these claims, and the theory of a connection between the aurora and zodiacal light was finally abandoned.

Quetelet gave careful attention to an enquiry as to whether there was any relationship between the periodicity of aurora and shooting stars (380). He admitted that auroras show a seasonal effect that is less evident for shooting stars but rested his argument in favor of some affiliation between the phenomena on the observation that of 28 unusual exhibitions of shooting stars between 1830 and 1841, 15 were accompanied or preceded by auroras. Baumhauer carried the argument further by point-

Sun, Weather, and Climate

The following is an abstract summary of a recent lecture presented by John A. Eddy at a national scientific meeting of the American Geophysical Union.

The question of whether the sun is responsible for significant changes in weather or climate has long intrigued science. In doubt is not whether the energy of sunlight drives the atmosphere or whether the sun is variable, for both are known to be true. At issue, rather, is whether known or unknown forms of solar variability are effective in altering day-to-day weather or the long-term trends of weather, known as climate.

Historically, the subject has been approached by searching for the signature of known periods of solar variability in weather records such as the incidence of flares, the 27-day period of solar rotation, or the 11-and 22-year periods of sunspot activity. This statistical approach has offered hints of possible connections but not always uniformly convincing ones. Moreover, our modern understanding of the complexities of both the sun and the atmosphere make simple connections unlikely. The related search for physical mechanisms of connections has been constrained by the lack of clearly defined statistical relationships and by the physical problem of coupling solar-induced changes at the top of the atmosphere to the much denser and more energetic troposphere.

In recent years, several new findings have given somewhat clearer indications of real sun-weather and sun-climate effects and at the same time have sharpened the limits of physical explanation. These include a suggested relationship between passages of solar magnetic sector boundaries and upper atmospheric circulation and a more clearly defined correlation between periods of extended drought on the High Plains of the United States and alternate minima of the 11-year solar cycle and a possible connection between long-term climate and the envelope of the annual mean sunspot number.

It seems important to clarify the physical constraints imposed on possible connections between the sun and the lower atmosphere, if their study is ever to advance beyond the realm of controversy, for the rewards for science and for mankind could be great, if significant connections are identified.

ing out that meteors are largely composed of nickel and magnetic iron which would remain suspended in the air parallel to the earth's magnetic field and would finally manifest itself as the aurora borealis when rendered luminous "by any cause." He even suggested a search of high-latitude regions for nickel on the ground which would, if found, be evidence for meteors being associated with aurora (43).

Boué at this time argued earnestly in favor of a connection between aurora and earthquakes (297). A table of monthly occurrences of earthquakes in the northern hemisphere showed a summer minimum but little detailed correlation with auroras. Boué attached importance to the fact that of 457 earthquakes recorded between 1837 and 1847, 47 corresponded to the day and 5 to the hour of a reported auroral occurrence Fifty others corresponded within 1 to 2 days.

It is clear that auroral physicists of the mid-nineteenth century could have benefited greatly from the still to be developed mathematical theory of statistics. It is curious that they did not take the lead of Dalton, who was far better versed in an understanding of meaningful probability when making comparisons of this type.

The Forgotten Explanation

A good story circulated late in the nineteenth century that illustrated the lack of understanding of the cause of the aurora (389). An honors student, when undergoing the torture of his *viva voce* (oral) examination, was unable to reply satisfactorily to any of the questions asked. "Come, sir," said the examiner, with the air of a man asking the simplest questions, "explain to me the cause of the aurora borealis." "Sir," said the unhappy aspirant for physics honors, "I could have explained it perfectly yesterday, but nervousness has, I think, made me lose my memory." "This is very unfortunate," said the examiner, "you are the only man who could have explained this mystery, and you have forgotten it!"

But this period did see the general acceptance of a relationship between auroras and activity on the sun. Mairan had first suggested a relation to sunspots, and Wolf investigated this possibility in great detail (512). He made elaborate monthly and annual comparisons between 1826 and 1843 of the number of solar spots seen on days when aurora were observed in comparison with the general average. Though the evidence was not decisive, he leaned toward Mairan's opinion and was emphatic about a connection between sunspots and the aurora of September 1, 1859. This was the event seen by Carrington (see page 81), who also strongly suggested a connection with the ensuing aurora, and today is usually credited with the discovery. The brilliant auroras of 1870 and 1871 were associated with large numbers of sunspots, and in 1873, Loomis (293) showed a periodicity of 10–11 years for large auroral displays, with times of maxima "corresponding quite remarkably with the maxima of solar spots."

This "solar connection" was the key to the development of our modern understanding of the aurora, though another 20 years had to pass before the discovery of cathode rays, or electrons, allowed the key to be turned. Appropriately, the door was opened in Norway.

Painting by Stephen Hamilton of the aurora during twilight from Labrador.

Earthquake Lights

It is of interest to discuss in a little more detail the claims of association between auroras and earthquakes. The reality of strange lights in the sky in association with earthquakes has been as elusive to prove as the reality of auroral sounds and for similar reasons—most testimony is from personal observations of untrained observers. There have been many reports of "earthquake lights" from earthquake-prone Japan (137), and an old Japanese lyric poem (117) suggests,

> The earth speaks softly
> To the mountain
> Which trembles
> And lights the sky.

Descriptions of these lights in Japan often compare the lights with auroral rays and glows, and the reported colors are similar to auroras.

Earthquake lights were reported from England as long ago as 1750, a year known for widespread seismic activity, when a Mr. Sedden at Warrington reported that at the moment of an earthquake (April 2) he saw "infinite numbers of rays of light, proceeding from all parts of the sky, to one point near the zenith." The rays were at first yellow, then blood red, and visible for 20 minutes (117).

From America in 1885, a Mr. Veeder reported aurora on October 7 and 8 and on November 2 and that corresponding to those times there was a renewal of earthquake activity in South Carolina (137). Lights associated with the Santa Rosa, California, earthquake of 1972 were described (137) as "auroral streamers diverging from a point on the horizon" (a seemingly impossible auroral configuration). The June 2, 1977, issue of *The New York Times* contained an article on the disastrous earthquakes that killed hundreds of thousands in China on July 28, 1976. Just before the first tremor, at 3:42 A.M., the sky over Tangshan lit up "like daylight," waking thousands who thought their room lights had been turned on. The multicolored lights, mainly white and red, were seen up to 200 miles away.

The statistical probability of simultaneous occurrence of auroras and earthquakes at mid-latitudes (assuming no causative relationship between the phenomena) is minute but necessarily finite and could account for rare "associations" between the two. More recent studies by Japanese seismologists include photographic evidence testifying to the reality of the phenomena, but the known physical properties of these lights are not consistent with auroral or ionospheric phenomena.

In modern times, earthquake lights must be examined carefully to ensure that the lights and flashes do not result from ruptured power lines, a probable source of a large fraction of the reports. However, many observations are accepted as genuine, and although the lights must be an atmospheric electrical phenomenon, the earthquake trigger mechanism remains to be discovered (137).

9
The Emergence of Norwegian Auroral Science

Kristian Birkeland once told his friend
Saeland—later professor of physics—
that the first thing he bought with
money he had earned himself as a boy,
was a magnet.

Olaf Devik
Kristian Birkeland as I Knew Him, 1967

Henri Becquerel 1852-1908
Professor at Museum of Natural History
of Paris and winner of Nobel Prize for
Physics in 1903

Adam F. Paulsen 1833-1907
Director, Danish Meteorological Institute

Let us now turn to the last decade of the nineteenth century and take up the story of the development of modern understanding of the aurora. The fact that magnetic disturbances and auroral displays concentrate around the poles had led to the belief by some that both phenomena were caused by high-speed corpuscles guided toward these regions by the earth's magnetic field. There was considerable difference of opinion on the origin of these charged particles. According to one hypothesis (50) suggested by Becquerel around 1878, the particles were shot off from the sun. He believed that hydrogen was ejected from sunspots and carried with it positively charged electricity. Goldstein (200) adopted a similar view in 1881 but believed the solar-terrestrial current to be composed of cathode rays. ("Cathode rays" was the name given by Goldstein to the flow of electricity across an evacuated glass tube. At that time, the nature of the material or particles that composed cathode rays was not known.) Another hypothesis suggested by Paulsen (365) in 1893 was that the particles were produced in the earth's upper atmosphere by intense ultraviolet radiation associated with solar flares.

It was at this time that the great Norwegian physicist, Kristian Birkeland, developed a passionate interest in aurora, an interest to which he devoted most of his life. His first auroral expedition to northern Norway in 1897 had as its main aims the determination of auroral heights and the checking on reports of very low auroras. The expedition was a failure, chiefly because of lack of experience in polar environments. In fact, Birkeland very nearly lost his life (56) when he and his party were ascending Beskades mountain in northern Norway (with reindeer and traps) and were caught in a severe blizzard. The second expedition, in 1899–1900, was more successful, and Birkeland obtained the first auroral photographs from separate sites (3.4-kilometer base line) from which it was possible to determine auroral heights (56).

However, it was his third expedition, in 1902–1903, that produced the first comprehensive data base for the study of magnetic perturbations and auroras. With financing of 20,000 krone from the government, 6,000 krone each from three interested friends, and 30,000 krone of his own, Birkeland established four arctic stations that straddled the auroral zone (Bossekop in northern Norway, Iceland, Spitzbergen, and Novaya Zemlya).

Analysis of simultaneous magnetic data from these stations led Birkeland to the conclusion that large currents flowed along the magnetic field lines during aurora (56). (This early work by Birkeland was neglected for more than half a century, and only during the last decade have auroral physicists again realized the importance of these "field-aligned currents." Study of such current systems has been one of the most active fields of auroral and space physics in the 1970's.)

Birkeland also became convinced of an association between auroral and meteorologic phenomena. In particular, he claimed that cirrus clouds formed parallel to auroral arcs and vice versa, implying that both were caused by the same beams of corpuscular rays. He suggested that the clouds resulted from "water vapour brought to condensation by the ions formed by the impact of negative (corpuscular) rays" that had their origin at the sun (56).

But Birkeland is not remembered as a great auroral observer, as it was his theoretical interests that guided him to his most inspired research on the aurora. The new research on cathode rays and the discovery of X rays

Northern lights over Norway from a nineteenth century engraving.

Birkeland with some of his auroral instruments in northern Norway, about 1900.

Kristian Birkeland 1867-1917
Professor of Physics at the
University of Oslo

in the early 1890's greatly excited Birkeland. Sir William Crookes had demonstrated that cathode rays were deflected by magnetic fields and in 1896, Birkeland proposed that auroras were caused by a beam of cathode rays emitted by the sun; those rays reaching the vicinity of the earth would be profoundly affected by the earth's magnetic field and guided to the high-latitude regions to create the aurora.

The following year, the great British physicist J. J. Thomson showed that cathode rays were composed of tiny negatively charged particles much smaller than hydrogen atoms and later called "electrons." Birkeland expanded his ideas to suggest that most of his solar electrons would be emitted from sunspots, and indeed it was well established that the 11-year period in sunspot activity was connected with variations of auroras and magnetic storms. To verify his theory, Birkeland first performed a special study of sunspots. Next, he organized arctic expeditions to acquire auroral and magnetic observations. And finally, he started model experiments in the laboratory.

Birkeland is best remembered for his long series of remarkable model experiments, in which he was ably assisted by his friend, Mr. Dietrichsen, a retired headmaster. The experiments consisted of projecting electrons in an evacuated chamber toward a sphere in which there was an electric coil, thus providing the sphere with an imitation of the earth's dipole

Birkeland with his terella apparatus, about 1910. A beam of cathode rays, or electrons, was directed at the magnetized sphere in the vacuum chamber. His assistant in the picture is Karl Devik.

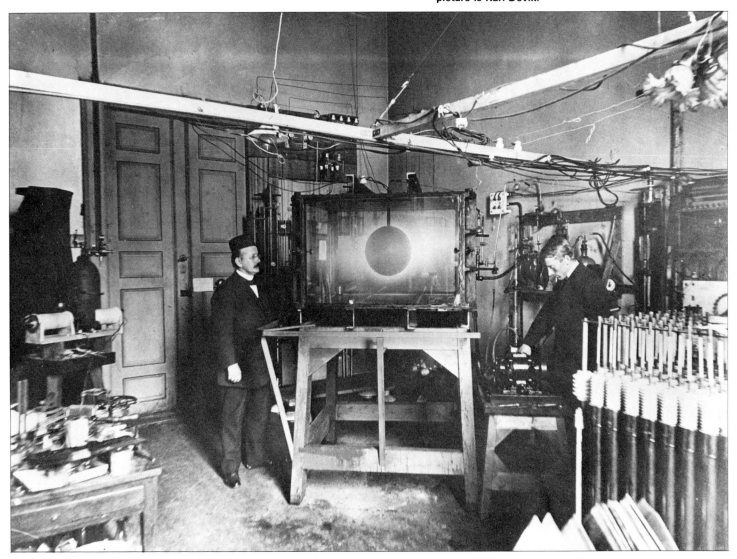

Three of Birkeland's photographs of precipitation regions on the terella.

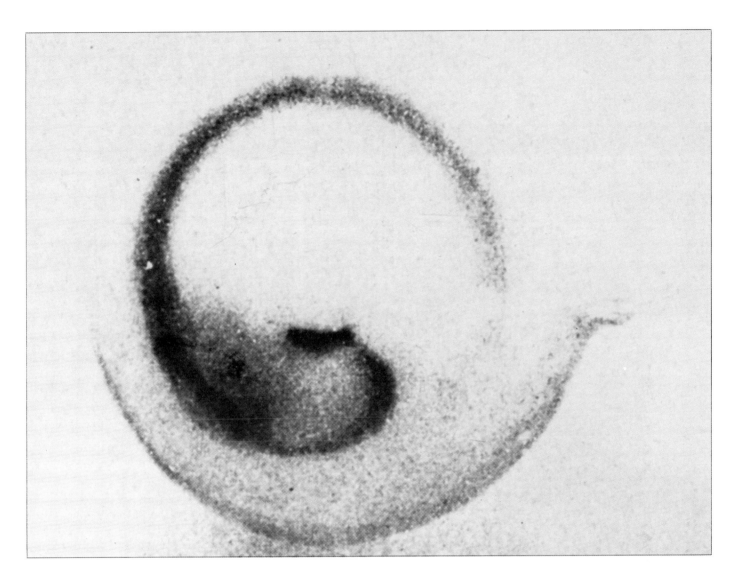

A polar view, showing electrons hitting the sphere in a zone suggestive of the auroral oval.

An equatorial view, showing electrons hitting the sphere in two zones in the northern and southern hemispheres, similar in configuration to the auroral zones.

Another of Birkeland's remarkable terella photographs, simulating the auroral ovals above the poles.

Sir Arthur Schuster 1851-1934
English physicist

Svante Arrhénius 1859-1927
Swedish physicist

field. By coating this "terrella" with fluorescent paint he was able to see where the electrons impinged on it. In his amply illustrated publications between 1900 and 1913, he was able to reproduce many features of the known auroral distributions. He was absolutely convinced of the correctness of his auroral theory even in the face of criticism by Schuster (410) on the grounds that a beam of electrons from the sun could not hold together against their mutual electrostatic repulsion. Birkeland's reply was that the electrons must travel at near the speed of light. (Though this was inconsistent with the known time delay of about 1 day between great solar flares and magnetic activity at the earth.)

At about this time, Arrhénius, in Sweden, was promoting very similar ideas as to the cause of the aurora and was presumably strongly influenced by Birkeland's work. Arrhénius described how negative particles from the sun would arrive more thickly over the equatorial regions of the earth, which were most directly exposed. But long before reaching atmospheric heights, they would be caught by the magnetic field and forced to spiral around it and travel north and south along the field lines. Soon the particles would be forced downward toward the magnetic poles and the lower denser regions of the atmosphere and begin to give out the "darting and shifting light of the cathode ray" (31).

Birkeland was a dynamic and enthusiastic physicist and inventor, whose interests led him in many directions. The technical museum in Oslo houses his electromagnetic gun, in which the projectile was a moving switch that short-circuited coils behind it to maintain a driving impulse. Birkeland's claims of a possible range of 60 miles brought representatives of Krupp from Germany for a demonstration. Birkeland recounted the story (140) as follows:

> . . . Then I turned down the handle. In a flash came a deafening and blinding row from an electric short circuit arc of 3000 amperes, with the flash of a long roaring flame out of the barrel. It was the most dramatic moment in my life; with a single shot I shot my shares exchange from 300 down to zero. However, the projectile hit the centre of the target!

But Birkeland's observation of this unexpected arc in a magnetic field led to his invention of the electromagnetic furnace for the production of nitrogen fertilizers, which became one of Norway's most important industries. Birkeland used financial returns from his more practical inventions to finance his first passion of auroral research. He also experimented with the radio telephone, tried to raise funds to work out methods of utilizing atomic energy, and built a demonstration to show how rocket ships in interplanetary space could obtain propulsion from cathode rays. His interest in the zodiacal light led him to establish an observatory in Egypt, where he was isolated from Norway after the outbreak of World War I. In 1917 he resolved to try to return to Norway via Japan and Siberia but died in Tokyo at age 50.

Birkeland's enthusiasm for the terrella models of the auroral mechanism had inspired a young professor in Oslo, Carl Störmer, to study mathematically the motion of charged particles in the magnetic field of a dipole, and this became Störmer's consuming interest for 50 years. Birkeland was delighted with Störmer's acceptance and mathematical development of his ideas and with the many apparent confirmations of experiments and theory. Without the aid of electronic calculators, Störmer laboriously computed detailed trajectories of

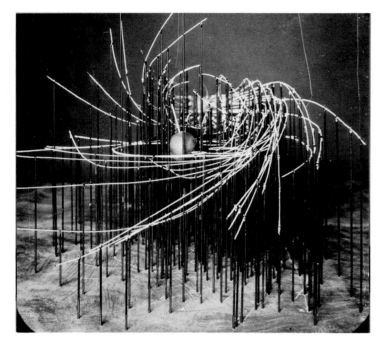

One of the wire models constructed by Störmer to display his calculated particle trajectories in the earth's magnetic field. The various trajectories (white wire) are supported by umbrella ribs.

Carl Störmer 1874-1957
Professor of Mathematics at the
University of Oslo

charged particles in the dipole field and built elaborate wire models to demonstrate the results. An amusing story concerns his use of the ribs from umbrellas to support his wire trajectories. The local umbrella manufacturer who was his source of these ribs became suspicious that the professor was developing a new improved umbrella at the university and so refused to supply any more ribs until the nature of their use was satisfactorily explained.

In 1907, Störmer published a paper (444) which created no stir at the time, but in the International Geophysical Year (see Chapter 13), exactly 50 years later, it suddenly became of historic importance. In it he set forth his calculations on the trajectories of charged particles in the two-pole magnetic field of a dipole, such as that of the earth. He showed the particles spiraled around magnetic field lines in tighter and tighter circles as they approached a pole, eventually "mirroring" over one pole and moving away from the pole again. Thus a particle would mirror over one pole, spiral back, and mirror over the opposite pole ad infinitum. This flying back and forth between mirror points is called "trapping" within the magnetic field.

Störmer himself did not attach too much significance to this result, as his chief interest was in calculating the convoluted trajectories followed by particles between the sun and the earth's atmosphere which might explain aurora. It turned out that Störmer's theoretical work applied to much faster or more energetic particles than we now know are appropriate for auroral excitation, but he had the satisfaction of finding his work had an important bearing on the theory of cosmic rays (very fast charged particles from outer space), a phenomenon unthought of when he began his calculations. And his results on trapping were of vital importance in understanding the vast belts of trapped Van Allen radiation that were not to be discovered for another half century.

The work of Störmer at his desk at the University of Christiania was a marvelous demonstration of the capability of man's intellect—he was calculating the behavior of particles too small to be seen in a region too remote to visit.

Störmer also became a great auroral observer, and from 1910 onward he took more than 40,000 auroral photographs, more than 9,000 sets be-

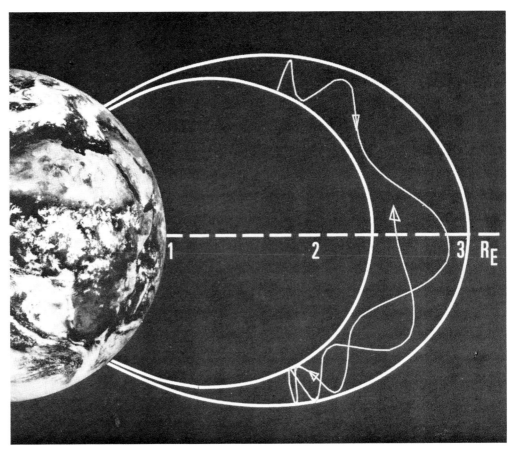

Störmer showed that charged particles spiral around the magnetic field and are reflected (or *mirror*) at low altitudes in each hemisphere, thus becoming *trapped* in the magnetic field. The *bounce time* from one hemisphere to the other is about 1 second.

ing taken simultaneously from two stations. Triangulation on these photographs clearly established the range of auroral heights, which he showed averaged 65–80 miles but could extend as low as 40 miles and as high as 600 miles (153).

Störmer's interest in photography was not confined to the aurora, and he was probably the most active candid photographer in Oslo in his day. The University of Oslo library has 40 volumes of his photographs of people and scenes around the city. In particular, Störmer enjoyed photographing young girls, and it is told that he walked around Oslo in the spring with a camera beneath his topcoat, in which he had cut a hole for the lens to protrude; in this way many of the young beauties of Oslo were unsuspectingly photographed, their latent images to mix with those of the aurora in the university darkroom.

The good custom of scientists is to appreciate the true discoveries and enlightened stimulus provided by great men, rather than unfairly use hindsight to dwell on any of their misconceptions, and from this view we must see Birkeland and Störmer as the truly great pioneers of modern auroral research.

Störmer poses with his auroral camera in Bossekop in 1910. The clock on the tripod (left) is
used to measure the time exposure. Taking notes (seated) is Olaf Birkeland.
(no relation to Kristian Birkeland).

A mass plot of many hundred's of Störmer's measurements of auroral height (obtained on his 1913 Bossekop expedition) shows the typical height of aurora is about 105 km.

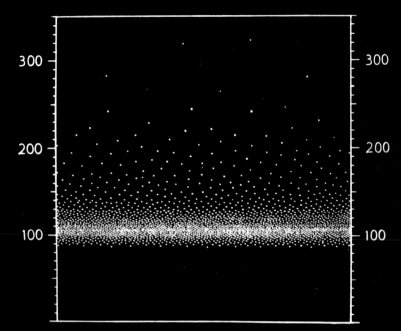

Working for Störmer at Tromso Observatory in northern Norway, this observer uses multiple cameras for good time resolution spectra of auroral development (about 1947).

Störmer meticulously kept all his scientific
records, correspondence, and drafts of
scientific papers. This painting was found
among these effects and is thought to have
been drawn by Störmer himself.

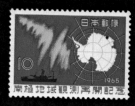

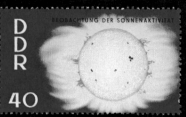

The aurora as featured on postage stamps from many countries.

10
Aurora Australis

And when, at last, in wearied beam,
Day faintly yielded to a star,
Austral Aurora's purple gleam,
To the horizon flashed afar;
And round the glowing zenith rolled,
In streaks of ruby and of gold.

J. C. Palmer
Thulia: A Tale of the Antarctic, 1843

AURORA AUSTRALIS

1908 – 09

**Title page from the first book printed on the antarctic continent by the men of Shackleton's 1907–1909 Expedition.
Drawing by George Marston.**

The story of the aurora australis is a very brief one compared to that of the aurora borealis. Although the southern lights have danced in antarctic skies for just as long as their northern counterpart has in the arctic, they have played to an empty theatre. The region of the southern auroral zone is located over the least traveled ocean of the earth and over the only uninhabited continent.

Until Europeans began to explore the southern oceans in the seventeenth century, probably the only humans to see the aurora australis were the native inhabitants of southern Australia and New Zealand, Tierra del Fuego, South Africa, and some of the Pacific islands. These people would only see the phenomenon on rare occasions when large magnetic storms resulted in mid-latitude auroras, and hence the event would probably be regarded fearfully. There are no written records from these cultures, but New Zealand Maori legend tells us that the aurora australis is a great fire lighted by their ancestors whose canoes had drifted far south to the Antarctic Ocean (34). Aborigines in Australia have not inhabited Kangaroo Island off the southern coast for 7,000 years; the spirits of the dead live there, and the lights of their campfires are sometimes seen flickering over the island (467).

Mairan, the author of the first textbook on aurora, seems to have been the first European to draw attention to the probable existence of auroras in the southern hemisphere. In 1750, he addressed a note of enquiry to Don Ulloa, the celebrated mariner of southern oceans. Ulloa replied (306) that while rounding Cape Horn in March and April of 1745, he often "saw a perceptible illumination . . . which had altogether the appearance of the polar lights so well known to him in the northern hemisphere." He added that such sightings might be expected to be rare, as the antarctic seas were seldom visited except during the summer, when the strong twilight would prevent auroral observations. This appears to be the first certain observation of aurora in the southern hemisphere made known in Europe, though an aurora was reported in Chile as early as 1640 (21).

A number of writers have (incorrectly) attributed the first sighting of the aurora australis to Captain Cook on his 1773 voyage on the *HMS Resolution*. Cook in fact had observed the southern lights during his first voyage of discovery to Australia on the *Endeavor* in 1770. The journal of Sir Joseph Banks (39) (botanist on the expedition) for September 16, 1770, states that

> About 10 o'clock a Phaenomenon appeared in the heavens in many things resembling the Aurora Borealis but differing materially in others: it consisted of a dull reddish light reaching in height about 20 degrees above the horizon: its extent was very different at different times but never less than 8 or 10 points of the compass. Through and out of this passed rays of a brighter colored light tending directly upwards; these appeared and vanished nearly in the same time as those of the Aurora Borealis, but were entirely without that trembling or vibratory motion observed in that Phaenomenon. The body of it bore from the ship SSE: it lasted as bright as ever till near 12 when I went down to sleep but how much longer I cannot tell.

This must have been a great aurora, as it was observed off Timor, just 10° south of the equator.

The first of six observations of aurora australis on the 1773 *Resolution* expedition was reported by Mr. Foster as follows (438):

Don Ulloa 1716-1795
Spanish mathematician and
naval officer

James Cook 1728-1779
English navigator and explorer

Sir Joseph Banks 1742-1820
Botanist on Cook's 1770 voyage

George Foster (right) and his father on
Cook's 1773 voyage.

Thaddeus Bellinghausen
1779-1852
Russian antarctic explorer

Charles Wilkes 1798-1877
American rear admiral and explorer

A beautiful phenomenon was observed during the preceding night, which appeared again this and several following nights. It consisted of long columns of a clear white light, shooting up from the horizon to the eastward, almost to the zenith, and gradually spreading on the whole southern part of the sky. These columns sometimes were bent sideways at their upper extremity, and though in most respects similar to the northern lights (aurora borealis) of our hemisphere, yet differed from them, in being always of a whitish colour, whereas ours assume various tints, especially those of a fiery, and purple hue. The stars were sometimes hid by, and sometimes faintly to be seen through the substance of these southern lights, but never had the fiery appearance sometimes seen in Sweden.

Subsequent voyages to the antarctic seas pushed farther south than Cook's had, and descriptions of auroras can usually be found in the logs of the expeditions. The Russian expedition under Captain Bellingshausen (1819–1821) reported a sighting on February 12, 1820, which was not recognized as an aurora. This is surprising, as they would surely have seen auroras in their native Russia. Bellinghausen's log (47) reads

At midnight we observed towards the south-west on the horizon a faint brightness resembling dawn and extending vertically almost 5°. As we proceeded southward this light extended further upwards. I concluded that it came from large masses of ice; however, as the day broke the light paled, and when the sun rose there was only white, very dense, clouds to be seen; there was no ice in that position. We had not, so far, observed any phenomenon of this kind.

A human touch is noted in the log of John Biscoe, captain of the brig *Tula,* who hints that the aurora was a nuisance because it distracted his men from urgent work. On March 3, 1831, at 65.7°S, he commented (535)

At the same time, nearly the whole night, the Aurora Australis showed the most brilliant appearance, at times rolling itself over our heads in beautiful columns, then as suddenly forming itself as the unrolled fringes of a curtain, and again suddenly shooting to the form of a serpent, and at times appearing not many yards above us; it was decidely transacted in our own atmosphere, and was without exception the grandest phenomenon of nature of its kind I have ever witnessed. At this time we were completely beset with broken ice, and although the vessels were in considerable danger in running through it with a smart breeze, which had now sprung up, I could hardly restrain the people from looking at the Aurora Australis instead of the vessel's course.

Probably the first picture of the aurora australis was the engraving published in the *Narrative of the United States Exploring Expedition* of 1838–1842 (under Commander Charles Wilkes). He described (503) a "splendid display extending all around the northern horizon . . . The spurs or brushes of light frequently reached the zenith, converging to a point near it." This aurora was described in poetry by J. C. Palmer (see title page of this chapter).

Many countries sent scientific expeditions to the Antarctic around the turn of the century, and voluminous reports of auroral observations resulted, Scott's expedition of 1901–1904 based at McMurdo Sound (the present staging area of U.S. antarctic activities) resulted in a number of beautiful watercolors of the aurora by E. A. Wilson. Wilson's descriptions of his paintings (505) express the frustration of artists anywhere in trying to capture the vibrant beauty of an auroral display (see page 253).

Captain Scott penned a fanciful description (412) of an antarctic aurora

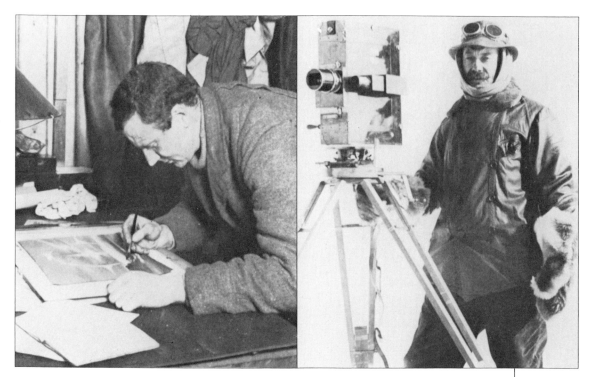

during preparation for his ill-fated trip to the south pole. On May 21, 1911, he wrote

> The green ghostly light seems suddenly to spring to life with rosy blushes. There is infinite suggestion in this phenomenon, and in that lies its charm; the suggestion of life, form, colour, and movement never less than of mystic signs and portents—the inspiration of the gods—wholly spiritual—divine signalling. Remindful of superstition, provocative of imagination. Might not the inhabitants of some other world (Mars) controlling mighty forces thus surround our globe with fiery symbols, a golden writing which we have not the key to decipher?

Scientific personnel on Scott's last expedition attempted to photograph the aurora but with no success. This puzzled them because they knew of Professor Störmer's success in Norway and had made notes on his methods. Ponting, who was attempting the photography, claimed to have faster lenses and plates than did Störmer but achieved no success even with very long exposures (412).

The most extensive auroral investigation on these early antarctic expeditions was carried out by Douglas Mawson on Shackleton's 1907–1909 expedition (316). He studied daily variations, monthly variations, and the relation to the magnetic meridian. He believed that local topography affected auroral shape, with the aurora deflecting to rise higher above mountains. Mawson also suspected some relation with local meteorology, though he felt more extended observations would be needed to derive the details of such an association. He theorized that the aurora resulted from currents flowing at high altitudes between the daylight and dark hemispheres in such a way as to neutralize a potential difference that should exist because of solar radiation ionizing the sunlit atmosphere.

It appears that the first successful photograph of the aurora australis was obtained by Mawson (316) from Cape Royal (Shackleton's winter quarters). He used 10-minute exposures to obtain impressions on his plates, "but as the curtains had altered their shape during the interval, the results were of little value." Mawson improved his equipment by the

Robert Falcon Scott 1868–1912
English antarctic explorer

Aurora australis by E. A. Wilson.

time he headed the Australasian Antarctic Expedition of 1911–1914 when he succeeded in obtaining auroral photographs with exposures as short as 10 seconds. The scientific reports of the expedition (316) also contain numerous sketches of various auroral types.

The remoteness and harsh environment of Antarctica has forced southern hemisphere research on aurora to take a back seat to the ongoing vigorous investigations of the northern lights. It was not until 1939 that sufficient observations had been gathered to allow the shape of the southern auroral zone to be determined (496) (whereas the northern zone had been mapped out 79 years earlier).

The increase in antarctic research stations in association with the International Geophysical Year led to more auroral research, but logistic problems have always been formidable for scientists wishing to work in this region. Most auroral questions can be equally well pursued in the northern hemisphere.

There are, however, some problems that require antarctic auroral observations. One of these is the question of auroral conjugacy. Conjugacy in this context means the occurrence of auroras simultaneously at both ends of the same magnetic field line. Mairan had suggested the possibility in 1733, and it was common knowledge in the nineteenth century that great auroras occurred in both hemispheres on the same days. After the general acceptance of the corpuscular nature of the auroral excitation and given that these charged particles would be guided by the earth's magnetic field, it was suggested that all auroras in the northern hemisphere would have a simultaneous counterpart in the southern hemisphere. But the geometry of the situation and the remoteness of Antarctica made experimental confirmation difficult. During northern hemisphere winter, antarctic skies are sunlit 24 hours a day and vice versa. And during the spring and fall the problem is to find two suitable observing stations (located on land masses and situated on the same magnetic field line). The solution was to mount cameras on aircraft and coordinate flight paths just south of Alaska and well south of New Zealand. These experiments (48) in 1968 demonstrated the expected conjugacy, and in fact surprised most scientists with the detailed similarities of conjugate aurora, both in shape and in intensity (see photographs on page 152). The southern lights can be regarded as a "mirror image" of the northern lights with an imaginary "giant mirror" located well out in space above the earth's equator.

Another aspect of auroral research that is uniquely suitable to antarctic investigation is that of dayside aurora. It has been recognized in recent years that the behavior of dayside aurora (i.e., auroras occurring near 1200 local time) relates very directly to one of the basic problems of auroral physics, that of the origin of the auroral particles. But to study auroras at 12 noon, one has to be at a suitable geomagnetic latitude and be at a high enough geographic latitude so that the sky is dark from horizon to horizon at midday. In the northern hemisphere this requirement is marginally satisfied at inaccessible locations in the Arctic Ocean near Spitzbergen (and then only for about 1 month near midwinter). But, South Pole, Antarctica is perfectly situated to study auroras near midday for a period of some 3 months every winter. Consequently, scientists at the Amundsen-Scott Base at the South Pole have been active in studying this very important sector of the auroral oval.

Sir Ernest Shackleton   1874-1922
English antarctic explorer

Sir Douglas Mawson   1882-1958
Australian antarctic explorer

These five plates show the aurora australis sketched on the British Antarctic Expedition of 1901–1904. All sketches are aurora as seen from winter quarters, the present site of the U.S. antarctic station at McMurdo.

Scientists would like to be able to proceed one step further and simultaneously study auroras near midday and midnight. This is not feasible from ground stations in the northern hemisphere (where the midday aurora occurs over the Arctic Ocean) nor in the southern hemisphere (because when the midday aurora is over the south pole, the midnight aurora is over the southeast Indian Ocean). These important experiments must await the launching of a satellite into a suitable orbit, so that one can photograph the complete auroral oval from space.

Aurora borealis observed at Melbourne, Australia, on September 2, 1859, at 10:20 P.M.

From the *South Polar Times,* volume 3, June 1902.

Illustration from the book *The Heart of the Antarctic* by E. H. Shackleton, 1909.

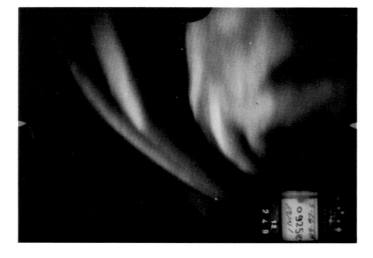
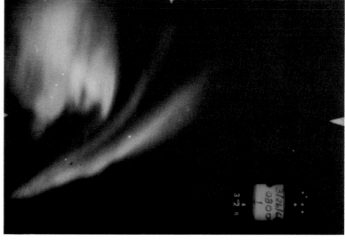

Photographs taken from jet aircraft flying under the same magnetic field line in the northern and southern hemispheres. This is a remarkable example of conjugacy of auroral forms, the southern hemisphere picture (between New Zealand and Antarctica) being a mirror image of the northern hemisphere picture (over Alaska).

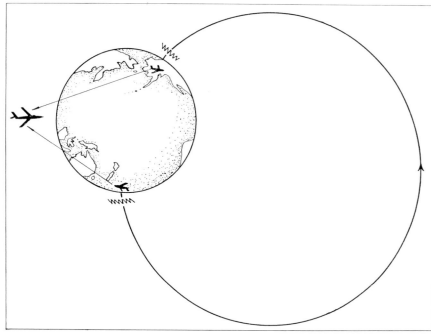

Two points near the earth's surface connected by a magnetic field line are said to be *magnetically conjugate*. To observe conjugate auroras, jet aircraft with cameras and instruments on board have flown out of Alaska and New Zealand.

Aurora at Scott-Amundsen Base, South Pole, Antarctica. The geodesic dome that covers the new American base is seen at the right.

11
Auroral Audibility

In sleep I heard the northern gleams;
The stars, they were among my dreams;
In rustling conflict through the skies,
I heard, I saw the flashes drive,
And yet they are upon my eyes,
And yet I am alive;
Before I see another day,
Oh let my body die away!

William Wordsworth
The Complaint of a Forsaken Indian Woman, 1798

The question of whether auroras make sounds is a very old one indeed. Although there have been regular reports of hissing, swishing, rustling, or crackling sounds over the past 300 years, the absence of any clear physical mechanism for producing the sounds has led to mixed feelings and skepticism on the part of scientists.

The earliest reference to auroral sounds (398) may date back to the Roman historian Tacitus (55–117 A.D.). In *Germania* (458) he discusses the Suiones, a tribe on the northern border of Germany:

> Beyond them is another sea, calm even to stagnation by which the circle of the earth is believed to be surrounded and confined; because the last gleam of the setting sun lingers till he rises again, and so brightly that it dims the stars. It is believed too that a sound is heard, that the forms of gods and rays from a head are seen.

An auroral interpretation of this passage is questionable, however, as Tacitus may have been referring to the midnight sun seen in the summertime at these high latitudes. A number of references in Norwegian sagas to the song of the Valkyries could be interpreted as referring to auroral sounds, as the aurora was explained as a flickering light from their armor as they rode through the northern skies. The careful treatment of auroral phenomena by the Norwegian father in *The King's Mirror* (about 1250 A.D.) does not touch upon the subject, though the next report in the literature does come from Norway. In his description of the aurora of December 1563, Absalon Pederssën Beyer (teacher and minister of Bergen) says (in diaries preserved at the Royal Libraries, Copenhagen) that

> . . . the sky grew red in the west and fire and flame darted back and forth so that a great noise was given off. I asked Christern Ulff what caused the sound, as I thought perhaps it came from the Alrichstadselff [a river near Bergen]. He replied, 'Don't you see it is in the sky? It was the clouds that ran rapidly back and forth.'

A significant percentage of residents of countries in or near the auroral zone believe that the aurora does make sounds, and the descriptions are remarkably similar: a faint rustling, hissing, swishing, or crackling, often compared by the Eskimos and Laplanders to the rustlings of a field of corn. An impression among many scientists that reports of auroral audibility come almost entirely from "natives" or less educated and easily misled observers is grossly in error and neglects the fact that the survival of many indigenous arctic peoples depends on accurate perception of their environment. Silverman and Tuan have recently examined the topic of auroral audibility in great depth (426) and have cataloged 197 reports of sounds associated with aurora over the past 250 years. These reports come from a wide range of people and locations and include observations by explorers, clergy, government officials, military men, physicians, and scientists, as well as observations from Eskimos, Indians, trappers, and the like.

Two questionnaire-type surveys on just this topic have been carried out, one by Tromholt (471) in Norway in 1885 and the other by Beals (44) in Canada in 1933. Approximately one in five people replied to the surveys, and these respondents would naturally be those most interested in the question and so the most highly motivated to answer. But even so, it is impressive that for both surveys, about 80% of the respondents claimed to have heard auroral sounds. On the other hand, many residents of the north and trained observers who have spent several years in the arctic are just as positive that the aurora does not make a sound.

154

A number of poets have given apt expression to auroral sounds. The quotation from Wordsworth at the beginning of this chapter is from a poem about the Indian custom of leaving the sick behind to perish alone in the snow if they cannot keep the pace of a journey. Wordsworth learned of this custom from Hearne's book (218) about his journey from Hudson Bay to the Northern Ocean (1795) in which mention is made of the "rustling and crackling noise" of the aurora "like the waving of a large flag in a fresh gale of wind." Wordsworth may also have been an acquaintance of John Dalton through his friend John Gough, who was one of Dalton's teachers; consequently, he may also have been influenced by Dalton's auroral studies (426).

The Scottish poet Robert Burns also speaks of auroral sounds:

The cauld blue north was streaming forth
Her lights, wi' hissing erie din
                    Libertie—A Vision

There are many other poetic quotations. Here are just a few.

You tink you hear da skruffel
of der lang green goons o sylk
                    T. A. Robertson
                    Nordern Lichts

If you listen, when weirdly the lights are streaming,
perhaps you may hear a whisper low.
                    T. A. Robertson
                    Northern Lights

And close ranked spears of gold and blue,
Thin scarlet and thin green,
Hurtled and crashed across the sphere
And hissed in sibling whisperings,
And died.
                    C. G. D. Robert
                    The Iceberg

They writhed like a brood of angry snakes,
  hissing and sulphur pale.
They rolled around with soundless sounds
  life softly bruised silk.
                    Robert W. Service
                    The Ballad of the Northern Lights

Here is an example of a poet who believes the aurora to be silent.

Rich as a sunset on the Norway coast, the sky,
  the islands, and the cliffs,
Or midnight's silent glowing northern lights unreachable
                    Walt Whitman
                    A Riddle Song

Attempts at the Geophysical Institute, University of Alaska, to record auroral sounds (by using highly sensitive microphones covering the complete frequency range of human hearing) have failed to provide any positive results. Yet many scientists remain convinced of the validity of reports on auroral sounds, rationalizing that they are rare and localized occurrences. A comprehensive statistical study of the subject by Silverman and Tuan showed that occurrences depend on the geomagnetic latitude of the observer and occur in a band near the auroral zone with a sharp cut-off to the north; reports are consistent with the known daily variation of the aurora and with the secular sunspot variation of aurora

William Wordsworth 1770–1850
English poet laureate

Samuel Hearne 1745–1792
English explorer of arctic Canada

Walt Whitman 1819–1892
American poet

and are more common during magnetically stormy conditions (426). These conclusions are hardly surprising, as by definition one needs an aurora to hear an aurorally associated sound. Of more interest are Silverman's and Tuan's conclusions that reports of auroral sound are definitely associated with overhead aurora (and are rare otherwise), are definitely associated with intense and rapidly moving forms (and are rare for weak arc forms), and seem to be localized within a range of just a few kilometers.

Another factor that influences the possibility of hearing auroral sound is the variation in sensitivity of the human ear between different observers. There are instances of groups of people all hearing sounds, and other examples of two people, side by side, disagreeing on the existence of sound. An interesting example of the latter, which actually makes a strong argument for the reality of auroral audibility, is provided by William Ogilvie (262), surveyor and explorer of the Yukon who was described as an exceptionally careful observer. On November 18, 1882, he noted that

> . . . One man, a member of my party was so positive [of hearing auroral sound] . . . I took him beyond all noise of the camp, blindfolded him and told him to let me know when he heard anything, while I watched the play of the streamers. At nearly every brilliant rush of the auroral light, he exclaimed: 'Don't you hear it?' All the time I was unconscious of any sensation of sound.

Silverman and Tuan considered a long list of suggested mechanisms for auroral sound. It has been argued that the close connection between the senses of hearing and seeing would lead to the expectation of sound in an auroral display, so that the brain might oblige; this would seem untenable, as observers note sounds only rarely and because of the connection with geophysical parameters. The latter objections also apply to the suggestion that tinnitus (a common hearing disorder of ringing in the ears) could be an explanation. A swishing sound is often heard when breathing air below −40°F (probably resulting from collisions of ice crystals formed in exhalations), and this has often been suggested as a mistaken source for auroral sound; but such "auroral sounds" should be

Paintings by William Crowder (opposite page) illustrated a 1947 *National Geographic* article on the aurora.

Engraved from a drawing made on the French Arctic Expedition to Bossekop, Finland 1838–1840. Seen to the east on January 19, 1839 at 9:27 P.M.

Auroral Noise and the Eskimo

A significant side light on the question of auroral noise comes from John A. Easley, Jr., who spent two winters (1943–1945) at River Clyde, Baffin Island, operating an ionosphere observatory during World War II. He learned enough of the Eskimo dialect to talk with some of the natives. In a letter (187) to Dr. Carl Gartlein at Cornell University, he said

One evening . . . during my second winter several Eskimos came into the house and remarked that the aurora was very bright. 'Aksaneealo! Eee-e, kowmayoalo! Peeooyoalo!' they said [Lots of aurora! Yes, very bright, very nice!]

Then one of them remarked suddenly, 'Neepeealo!' [very noisy], and the others grinned and agreed enthusiastically. I grabbed my parka and dived for the door. This was what I wanted to observe, a noisy aurora. But when I got outside and stopped to listen, I could not hear a sound.

The natives came pouring out of the house, laughing at my queer behavior, and I questioned them immediately about the meaning of their statement 'Neepeealo.'

'What noise did you hear?' I asked.

'Oh,' they said, 'there's no noise. The Eskimos say it's noisy when the aurora moves rapidly [Aksanil audlatidloalo]. They say that there must be noise up there, because it moves like a cloth in the wind!'

'Oh,' I said, 'some people say they hear the aurora.'

They laughed again, this time with that attitude so characteristic of them, that the white man will never learn much about the north.

Perhaps this indicates that some previous reports of noise associated with the aurora may have resulted from a misunderstanding of what Eskimos meant.

Illustration from a popular book on arctic expeditions (1877) showing Eskimos, their igloos, and the aurora borealis.

heard in the absence of auroras and not at warmer temperatures, which does not seem to be the case.

Direct transmission of sound from auroral heights can be ruled out immediately on two grounds: propagation attenuation would require a sound source at the aurora that is 1,000 times the energy required to generate the brightest auroras, and it would take any sound so generated several minutes to travel the 100 kilometers to ground level. More exotic possibilities, such as human perception of electromagnetic radiation and pressure effects from auroral electric fields or radio waves, are all rejected by Silverman and Tuan on firm theoretical grounds.

Only one mechanism seems acceptable to the physical scientist, and that is brush discharges caused by aurorally induced electric fields (426). It is known that intense auroras produce strong electric fields at ground level, and peaks as high as 10,000 volts per meter have been reported in association with very bright auroras. Though the details of how these voltages arise are not clear, it is evident that given the fact it happens, electrical discharges will occur from points where sharp voltage gradients exist. Such points may be trees, bushes, pine needles, etc., and the effect is increased at surface irregularities such as an isolated tree or a mountain. These discharges will give a rustling or hissing sound, somewhat dependent on wind conditions. Under completely still conditions, voltages could build up high enough to cause minute sparks, and this would produce a crackling sound.

If brush discharge is the explanation of auroral sounds, then one would not expect reports of sound from Antarctica (where no vegetation grows), and indeed this seems to be true. There are a few antarctic reports, but all seem to be explainable by freezing breath; typical is the following description by Amundsen (18) just before his dash for the South Pole:

> I was called out one very cold night by Johannsen to hear what Johannsen described as the crackling of the aurora australis. It was a very cold and very still night. I distinctly heard a very faint rythmically repeating rustling noise in the air. After a time I discovered this was due to the rapid freezing of the moisture from my breath, and the tiny tinkle made by the minute crystals as they slowly descended under gravity close to my face, sufficiently close for the ear just to catch the faint sound. There was no doubt that the rustling sound coincided with periods when I exhaled air.

The brush discharge explanation seems to fit all the reported characteristics of auroral sounds without straining one's credibility too far, so scientists should feel more comfortable about the subject. If verbal reports from impressionable humans are still viewed with suspicion, then perhaps descriptions of the behavior of dogs to the phenomenon might convince the skeptics. Gmelin (196) traveled in the northeastern districts of Siberia in the 1760's, where hunters described the auroral sounds they hear as *spolochi chodjat,* that is, "the raging host is passing." They reported that they were often overtaken by the northern lights as they pursued the white and blue foxes to the icy seas, and their dogs were so frightened that they lay obstinately on the ground until the noise passed. Others report that dogs have whined and turned in circles as the sound rises or immediately jumped up and commenced to growl (71).

The brush discharge mechanism also conveniently accounts for another curiosity of the aurora that has always met with complete skep-

Roald Amundsen 1872-1928
Norwegian polar explorer and first man
to reach the South Pole

Postage stamp issued by the USSR to commemorate Amundsen on the fiftieth anniversary of his death.

Aurora borealis observed at Nulato, Alaska, by F. Whymper on December 27, 1866.

160

ticism, that of auroral odor. There are perhaps only half a dozen reports (21) of smells in association with aurora, but though coming from different countries and different centuries, they agree in the description of a sulphur or ozone like odor. This is the same smell we often experience from worn kitchen appliances that contain an electric motor, and it results from the dissociation of oxygen molecules by spark breakdown and subsequent recombination to form ozone.

The question of auroral sounds remains unsolved, as was nicely summarized by C. R. Rust (162), an Alaskan old-timer:

Baron von Humboldt 1769-1859
German naturalist and explorer

> Aurora borealis, the scientists call them, but to us old sourdoughs they will always be the northern lights. Many nights I have stood outdoors almost freezing (and getting a kink in my neck) gazing upward almost by the hour when the display has been active and colorful. To me it is always a source of wonder and delight and at such times I wish I were a poet or had the magic of words to express my feelings. Often I just hurt from the awe and beauty and wondrous mystery of the lights . . . On very cold nights . . . they would hiss and crack . . . Scientists at our University of Alaska say the northern lights do not make any sounds, nor do they come down low. Many of us old-timers do not agree. I am no scientist. I can only try to describe what I have actually seen and heard many times during my fifty-two years in Alaska.

It seems today that even the Eskimos may be having second thoughts, as a group in western Alaska has said that, "things are not the same as they used to be, because in the early days the northern lights howled a great deal more than they do now." This comment is reminiscent of Humboldt's remark (242) in his famous *Cosmos* (1847), where he says, "Northern lights appear to have become less noisy since their occurrences have been more accurately recorded."

To prove scientifically that auroras sometimes generate a sound and that it is generated by brush discharge, is not an easy task. It would require sensitive audio equipment, measurements of electric fields, auroral photography, and careful monitoring of meteorological conditions. Other unexpected factors might enter into the conditions for the generation of auroral electric fields. And the phenomenon is extremely elusive, only rarely being reported for the brightest aurora. Consequently, a coordinated experiment designed solely to detect definitively auroral sounds would be unrealistic in terms of the time and money involved. And as it is regarded as a fringe phenomenon of questionable validity, it is not likely that the funding or effort will ever be devoted to the subject. But if one day the brush discharge explanation is proved to be true, those who had been convinced of the reality of the phenomenon will no doubt have mixed feelings: relief that their auditory system is to be trusted but disappointment that the source of sound is not, after all, high in the heavens.

"Same old ice, same old aurora borealis, same old everything!"

"Well, if it isn't the aurora borealis, I just hope it isn't another of those shopping centers."

Cartoons from *The New Yorker*.

Without You, Even The Aurora Borealis Seems Dull

The aurora featured on a greeting card.

12
The Contribution of the Spectroscopists

With its frigid brilliance, its blue-red sweeps
And gusts of great enkindlings, its polar green
The color of ice and fire and solitude.

Wallace Stevens
The Auroras of Autumn, 1949

Twinkle, twinkle little star,
I don't wonder what you are,
For by spectroscopic ken
I know that you are hydrogen.

D. Bush
Science and English Poetry, 1960

Sydney Chapman 1888-1970
English physicist, regarded as the
pioneer of modern solar-terrestrial
physics

Frederick A. Lindemann
(Viscount Cherwell) 1866-1957
Professor of physics at Oxford and
Science Advisor to Winston Churchill

Anders Jonas Angström 1814-1874
Swedish physicist

A. L. Wegener 1880-1930
German geologist, widely known for his
continental drift theory

In 1918, young Sydney Chapman, then at the Royal Observatory in Greenwich, published (98) a study of worldwide magnetic disturbances to which he added a theory based on a corpuscular stream of electrically charged particles from the sun of one sign only. Such a suggestion from Chapman was surprising, as he had met Birkeland and was familiar with Birkeland's ideas (56) and Schuster's criticism (410) (see page 134). Indeed Chapman was quickly challenged in 1919 by Lindemann (283) on the basis of the mutual electrostatic repulsion in such a stream (which would blow the stream apart before it ever reached the earth) and the implied build-up of an electric charge in the earth's atmosphere. Lindemann added the constructive suggestion that Chapman's theory of magnetic disturbances might be preserved in substance if the proposed corpuscular stream was replaced by a stream of charged particles of both signs in equal numbers, i.e., a neutral ionized stream, which we now call a "plasma." It was this idea and consequences of it that occupied Chapman and his co-workers (106) for much of the ensuing 50 years of his long and productive scientific career; and it was this idea that became central in our understanding of the marriage between sun and earth that produces those vibrant offsprings, the northern and southern lights.

But how was science to test these ideas of streams of charged particles between the sun and the earth? Four decades were to pass before rockets and satellites allowed direct measurements above the atmosphere. During this period the principal advances in experimental auroral physics were tied to increasing sophistication in the art of spectroscopy. This is the branch of science that accurately measures the wavelength (or color) of light. Study of such measurements (spectra) reveals a wealth of detail regarding the source. Chemical composition, pressure, temperature, and velocity are examples of properties that can be obtained from spectral analysis.

Angström, the Swedish physicist who pioneered. spectral studies, made the first crude measurements (22) of the auroral spectrum in 1866–1867. He reported the strongest yellow-green emission, which is responsible for the usual color of the aurora, and assigned a wavelength of 5567×10^{-8} centimeters, or 5567 Å (The unit of length of 1/100,000,000 of a centimeter is now called the angstrom.) Other weaker lines were soon reported in the red and blue regions of the spectrum, but no similarity could be established with known atmospheric gases. However, the fact that the aurora did exhibit a line spectrum proved it was self-luminous and not the result of reflected or diffracted light or of glowing macroscopic particles, as these latter sources would give continuous spectra. (An analogy would be the pure yellow light given by sodium street lamps in comparison with the white light, composed of all colors, emitted from an ordinary light globe with its hot tungsten filament.)

Over the next 30 years, the auroral spectrum was compared to the spectrum of almost every known element and to the spectra of the sun, comets, gaseous nebulae, and zodiacal light. Owing to the low resolution of spectroscopes of the period, many false identifications were made, and as late as 1915, a new physics text by Sir William Ramsey (382) stated that the auroral spectrum contained numerous lines, "all of which have been shown by Mr. Baly to be identical with strong lines in the spectrum of krypton." The German geologist Wegener argued that an unknown element in the sun's atmosphere was responsible for the strong green line, and he called this element "geocoronium" (491).

Spectroscopy came of age with the development of photography, which allowed permanent recording of spectra for later analysis. The association of major auroral emissions with atmospheric gases began in Norway in 1912–1913 with Vegard's work (482), leading to the identification of the auroral nitrogen bands in the blue and red regions of the spectrum. However, the elusive yellow-green line, the strongest line in the auroral spectrum, defied interpretation until accurate measurement of its wavelength as 5577 Å by Babcock (36) in 1923 finally allowed McLennan and Shrum (323) in Toronto to show that it resulted from a metastable transition of atomic oxygen. That statement needs a little elaboration: most light emitted by excited atoms results from so-called "allowed transitions"; that is, electrons in an atom are raised to higher-energy levels by some external source of energy, from which they almost immediately return to their original state and radiate the excess energy in the form of a burst of light. This process may typically take something like a millionth of a second or less. However, there are certain rules of atomic physics governing just how electrons may move between various energy states of the atom, and sometimes an electron can find itself raised to a certain energy state with nowhere to go, unless it breaks these magic rules of atomic physics. These are so-called "forbidden transitions," but they are not forbidden in an absolute sense; the statistical probability for the transition is very low so that the atom will remain in the energized state for a long time before it finally radiates its excess energy as light. Such an energized state is referred to as a metastable level, and in the case of the yellow-green light of atomic oxygen, the average residence time of the electron in that level is three fourths of a second.

Sir John McLennan, who with Shrum in Toronto in 1923 succeeded in identifying the strongest yellow-green line in the auroral spectrum as being a metastable transition from atomic oxygen.

This explains why the strongest of lines in the aurora had not been identified in laboratory spectra of oxygen. The lowest vacuums that can be obtained in the laboratory are such that atoms still collide with each other many times in ¾ second; such collisions remove the excess energy from the metastable atom ("collisional deactivation," or "quenching") without the emission of a light burst, the energy going into heat. However, in the atmosphere above some 60 miles, where the aurora occurs, there is less than one collision per second between atoms, so energized oxygen atoms have time to radiate the characteristic yellow-green line of the auroral spectrum. Auroral light coming from higher levels in the atmosphere is characterized by another atomic oxygen line becoming strongest at a wavelength of 6300 Å at the red end of the spectrum. This too is a metastable transition with a lifetime of 110 seconds, so it can only radiate above some 150 miles, where collisions between atoms occur less than once every 100 seconds.

With this spectral information we can now explain the various colors of the aurora as being related to height. In low auroras, below 60 miles, where even the green oxygen line is "quenched," the blue and red bands of the nitrogen spectrum dominate. Between 60 and 150 miles, the green oxygen line is strongest, and above 150 miles, the red oxygen line is most important. Thus provided the aurora is bright enough to exceed the color threshold of human vision (if not, it will appear white), the strongest light emissions can be in the blue, green, or red regions of the spectrum, and these of course are the three primary colors. Thus various combinations of light from different heights will mix to give the fantastic variety of subtle hues of blues, mauves, purples, and pinks that we see in the aurora.

AURORA SPECTRUM, SOLAR SPECTRUM, AND CANDLE SPECTRUM.

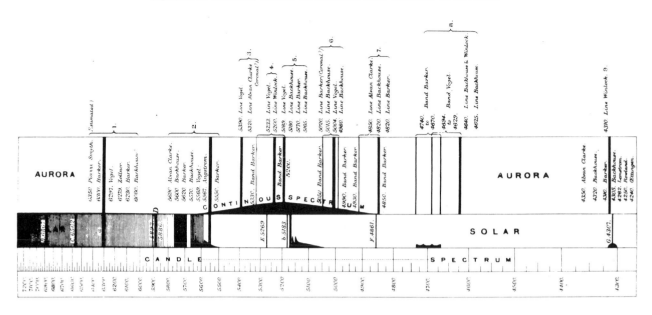

SPECTRUM OF THE AURORA.

The auroral spectrum was compared to the spectra of the sun and all the known gases to try to identify the source of the aurora.

The auroral spectrum (above) indicates the auroral lines measured up until about 1878 and a comparison with absorption lines in the solar spectrum (there is no correspondence).

The spectra (right) compares the auroral spectrum as it was known about 1885 with spectra from air, oxygen, nitrogen, and hydrogen. Again there is little correspondence, except for some of the hydrogen lines. This correspondence could not be real, however, as instruments at this time were not sensitive enough to detect the weak hydrogen lines in the auroral spectrum.

Expédition du cap Thordsen, tome II, 1er fasc   Planche XXIX

SPECTRE DE L'AURORE BORÉALE

COMPARÉ

AVEC CEUX DE L'AZOTE DE L'OXYGÈNE ET DE L'HYDROGÈNE

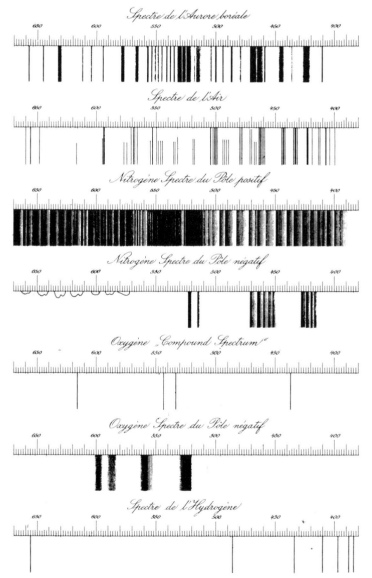

Spectre de l'Aurore boréale

Spectre de l'Air

Nitrogène Spectre du Pôle positif

Nitrogène Spectre du Pôle négatif

Oxygène "Compound Spectrum"

Oxygène Spectre du Pôle négatif

Spectre de l'Hydrogène

Central-Tryckeriet, Stockholm.

AURORÆ

AND THEIR

SPECTRA

Lars Vegard, the Norwegian auroral spectroscopist, made many important identifications of nitrogen, oxygen, and hydrogen emissions in the auroral spectrum. Here he is shown with his spectrograph about 1915.

But what causes the aurora to have such a range of heights? If one accepts the corpuscular theory whereby atmospheric atoms and molecules are energized by collisions with incoming fast particles, then an obvious answer is that faster particles will penetrate deeper into the atmosphere and cause lower aurora. This idea was known in the 1920's, but direct proof was lacking.

Auroral theory was waylaid for a few years beginning in 1928, when Hulburt revived Paulsen's ultraviolet light hypothesis (365) and attempted to build a complete theory of aurora and magnetic storms around the idea (241). He proposed that copious ionization of atmospheric constituents, particularly over equatorial regions, was produced by occasional intense blasts of ultraviolet radiation from the sun. By various processes the ions were assumed to ascend to great heights from where they could be guided by magnetic lines of force to polar regions to produce auroral phenomena. Chapman took exception to these ideas (99) and pointed out a fundamental difficulty, that the velocity of ions produced was not high enough to let them penetrate down to the level of auroral displays.

Spectroscopy came back into the picture in 1939, when Vegard made the important discovery (483) of the hydrogen lines in the auroral spectrum, but the true significance of this observation did not become apparent until 1948, when he reported (484) that the hydrogen lines were Doppler shifted. This discovery was confirmed by Gartlein (188) and Meinel (325) in 1950–1951, and so for the first time it had been proven that fast particles (hydrogen atoms) bombard the upper atmosphere during aurora. To explain what is meant by Doppler shift, we turn to the familiar example of the apparent change in pitch, or frequency, of a train's whistle as it approaches and passes an observer. As the train approaches, the whistle sounds high pitched, changing to a lower pitch as the train recedes. This familiar effect with sound waves is also true of light waves; if an energized atom is approaching the observer with a high velocity when it emits its excess energy as a burst of light, the light is measured to have a higher frequency (or shorter wavelength) than if the atom was moving away from the observer. Gartlein and Meinel noted that the hydrogen line was shifted about 10 Å toward the blue region of the spectrum if they looked directly up the magnetic field direction, whereas it was at its expected position if they looked perpendicular to the magnetic field. This proved that the hydrogen light was emitted by fast hydrogen atoms traveling down magnetic field lines toward the

E. O. Hulburt
American physicist
(Naval Research Laboratory)

A modern low-dispersion auroral spectrum shows the main auroral emissions in the visible region of the spectrum. The auroral spectrum is composed of *lines* and *bands*, as distinct from the *continuous* spectrum of the sun or hot incandescent bodies. Note that there are strong emissions in the blue, yellow-green, and red regions; these are effectively the primary colors, and mixtures in varying ratios result in the wide variety of hue and color of the aurora.

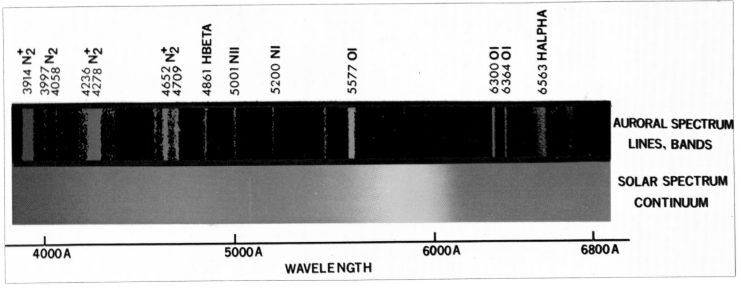

169

earth. The mechanism involves fast protons (ionized hydrogen atoms) being guided by the earth's magnetic field down to the atmosphere, where they collide with oxygen or nitrogen and capture an electron, forming an energized hydrogen atom. This atom quickly radiates the excess energy in the form of light, giving the Doppler-shifted hydrogen line.

This new discovery was hailed by many as finally solving the problem of the excitation mechanism of auroras, and for many years it was considered that protons, probably originating at the sun, were the particles responsible for the aurora. Further progress was halted by World War II, but after 1945 the rate of advance of auroral science was accelerated, partly due to advanced techniques developed during the war in rocketry, radio, and radar. Toward the end of the 1950's spectroscopists were coming to the conclusion that it was difficult to account for the auroral spectrum by proton excitation alone. High-altitude balloon and rocket flights between 1955 and 1958 detected X ray radiation below auroras (508); x-radiation is emitted by fast electrons as they pass close to an atomic nucleus, and this *bremsstrahlung*, or "braking" radiation, penetrates farther into the atmosphere than the electrons themselves. These measurements gave the first definite indication that fast electrons contribute to the auroral excitation.

So before the beginning of the International Geophysical Year of 1957–1958, the entry of particles of both signs into the polar atmosphere was known, but the share of the total energy carried by protons and electrons was not clear. The stage was thus set for more powerful rockets to take detectors into and above the aurora to resolve the question.

13
The International Geophysical Year

The satellite is a natural extension of rockets, which are natural extensions of planes and balloons, which are natural extensions of man's climbing trees and mountains in order to get up higher and thus have a better view.

James Van Allen
Time, May 4, 1959

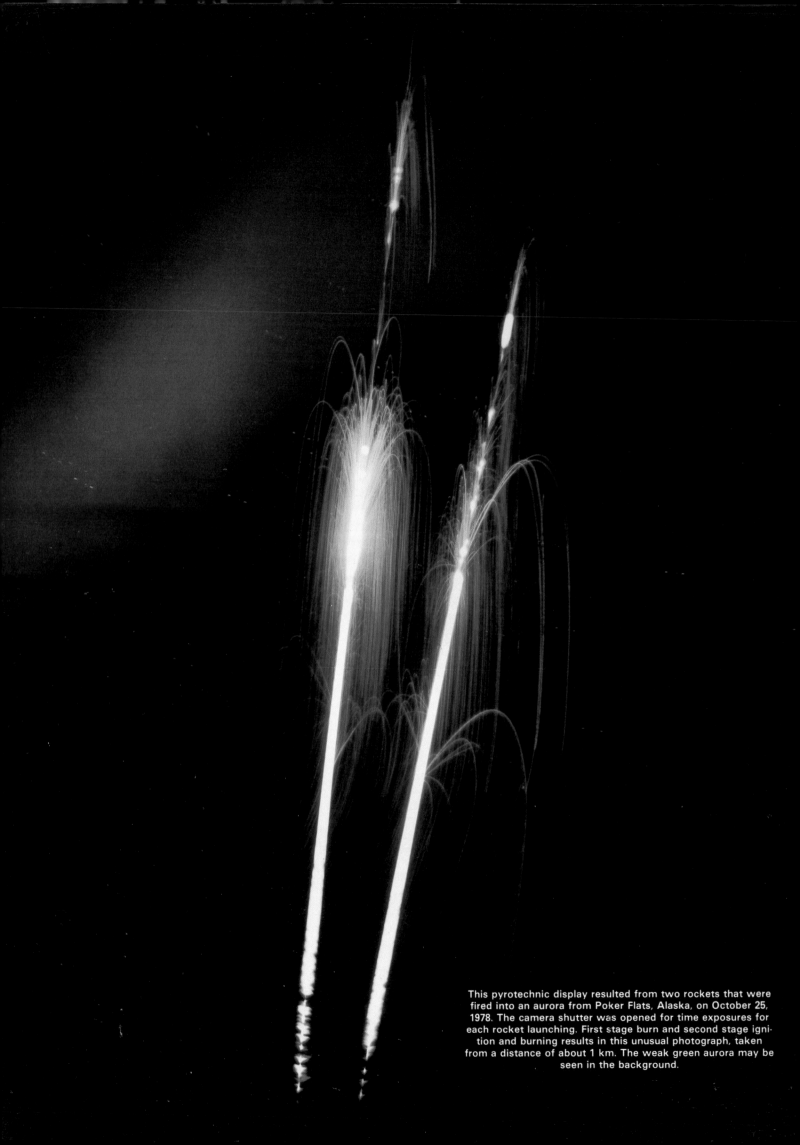

This pyrotechnic display resulted from two rockets that were fired into an aurora from Poker Flats, Alaska, on October 25, 1978. The camera shutter was opened for time exposures for each rocket launching. First stage burn and second stage ignition and burning results in this unusual photograph, taken from a distance of about 1 km. The weak green aurora may be seen in the background.

The International Geophysical Year (IGY) had its beginning at a dinner party in a Washington suburb at the home of James Van Allen on the evening of April 4, 1950, when it was suggested that another Polar Year should be organized in the forthcoming solar activity maximum of 1957–1958. Sydney Chapman was present, and it was he who shaped the broader notion of the International Geophysical Year. By accepting the position of President of the Special Committee for the IGY he committed himself to the years of planning, execution, analysis, and synthesis that became an unprecedented enterprise of scientific cooperation around the world.

Sydney Chapman 1888–1970
English physicist, regarded as the pioneer of modern solar-terrestrial physics

The IGY period heralded a basic change in the nature of auroral research. The challenge to the individual scientist working with basically inexpensive instruments, the adventure of research expeditions to polar regions, and the satisfaction of slowly unlocking the mysteries of nature's most elusive phenomenon were all to be overshadowed by the "brute force" approach of powerful rockets, complex satellites, and large group efforts. The aurora was to be caught up in the explosion of space research in the 1960's. Suddenly, man had the technology to make a determined effort to understand the magnetic and charged particle environment around the earth, the "magnetosphere." And the aurora, the only visible manifestation of magnetospheric processes, became a focus of interest for this new generation of space scientists.

The first rockets were launched into auroras early in the IGY by Meredith and colleagues (327) and by McIlwain (321) from Ft. Churchill in Canada. Particle detectors on board these rockets showed that the aurora was principally excited by energetic electrons bombarding the upper atmosphere. Typically, fluxes required for a moderate aurora are 1,000 million electrons per square centimeter every second, with average speeds of 20,000 kilometers per second. Protons are also present, and they are usually found in a much broader diffuse region which may or may not coincide with the electrons. The protons generate a subvisual aurora called the proton or hydrogen aurora; bright electron auroras can deposit 100–1,000 times the amount of energy per unit area that is precipitated in typical proton aurora. The two types of precipitation often coexist, bringing approximately equal amounts of energy into the atmosphere and generating a broad low-intensity diffuse aurora.

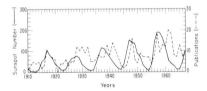

Yearly number of Chapman publications and yearly relative sunspot numbers, 1910–1967. Not only auroras follow the 11-year sunspot cycle. Sydney Chapman published more than 450 papers in his lifetime, and his interest in solar-terrestrial physics is clearly shown.

The IGY also saw an extensive network of all-sky cameras set up to study worldwide auroral morphology. All-sky cameras photograph the whole sky from horizon to horizon and had previously been used for cloud research; improvements from various countries led to the IGY network of 114 cameras in the Arctic and Antarctic; in addition, the Auroral Data Center at Cornell University received about 18,000 hourly reports a month from some 430 amateur sky watchers, including airline pilots, seafarers, science teachers, and the like; 130 U.S. Weather Bureau stations made hourly observations. The advent of electronic computers in the 1950's offered the hope of digesting the mountains of auroral data thus collected.

It was auroral observations that involved more amateur enthusiasts than any other IGY effort. Amateur observations provided a picture of auroral displays more complete than was possible ever before. It is easy enough for one man to look up and see an aurora but to combine hundreds of such sightings into a worldwide pattern was the unique contribution of the IGY. Sometimes the picture that emerged from the

assembled data was awesome. At one time Carl Gartlein, head of the IGY Auroral Data Center, reported (450) that

> There was a wall of light as long as the U.S. is wide, over 100 miles tall, with its bottom 60 miles from the ground, moving south at 700 miles per hour.

The all-sky camera data proved to be the most useful, however, and analysis of the hundreds of thousands of photographs taken at 1-minute intervals throughout the dark hours revealed two important large-scale features of the aurora. In 1963, Feldstein in the USSR showed that auroras appear along an oval-shaped band rather than along the auroral zone (170). Recall that the auroral zone is that statistical region of maximum probability of seeing auroras and is a circle located near 67° magnetic latitude. The instantaneous distribution of auroras described by Feldstein is oval shaped and is located near 67° magnetic latitude at local midnight but at about 78° latitude near local midday (for quiet magnetic conditions). Because auroras are brightest and therefore most readily observed near midnight, the statistical auroral zone is, in effect, the locus of this brightest midnight part of the oval.

The other large-scale feature, discovered by Akasofu (4) in 1964 from IGY camera data, was a sequence of systematic and characteristic auroral displays which he called the auroral "substorm." Quiet auroras which lie along the auroral oval are intermittently activated. It is at these times that we see the bright, active, and spectacular auroral displays. In general, this activation originates near midnight, though its effects are rapidly seen at earlier and later local times. The substorm reaches its peak in a rather short time (~10 minutes) and gradually subsides. During very quiet magnetic conditions a substorm might occur only once a day; during moderately active periods, substorms occur every few hours.

But the IGY is probably best remembered as the beginning of the Space Age. Scientists were using captured German V-2 rockets as early as 1946 to explore the upper atmosphere, but we were still living in an essentially two-dimensional world. We could observe meteor trails near 100 kilometers and auroras to hundreds of kilometers, but what went on in the upper atmosphere and the ionosphere beyond was unknown.

The satellite era began October 4, 1957, when *Sputnik*, the Russian word for satellite, was dramatically added to all languages. The launch of *Sputnik I* produced considerable political reaction and propaganda around the world, especially in the United States, where the public and press demanded to know why the United States was behind in this "space competition."

The United States had been working on satellites for 2 years but was a long way from a launch date. The idea of satellites was not especially new; an Air Force commissioned report (383) in 1946 entitled *Preliminary Design of an Experimental World-Circling Spaceship* concluded that

> A satellite vehicle with appropriate instrumentation can be expected to be one of the most potent scientific tools of the Twentieth Century. The achievement of a satellite craft by the United States would inflame the imagination of mankind and would probably produce repercussions in the world comparable to the explosion of the atomic bomb.

But funding problems and competition between the Army, Navy, and

Yasha Feldstein
Russian auroral physicist

Syun Akasofu
Professor at Geophysical Institute
University of Alaska

(Opposite, top) Auroral studies have benefited greatly from the cooperation of amateur observers, especially during the IGY. Here amateur astronomers of the Milwaukee Astronomical Society are taking photographs and measuring auroral brightness (1957).

(Opposite) An early all-sky camera (1947). The convex mirror, mounted on the box, reflects the entire sky, and the camera, aimed at the mirror, photographs the reflected image of the aurora. This allowed the full upper hemisphere of the sky to be photographed. Carl Gartlein is changing a date label while his children watch.

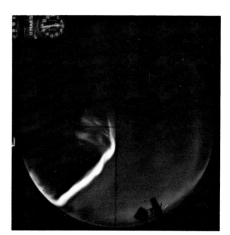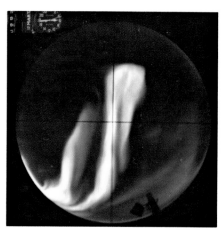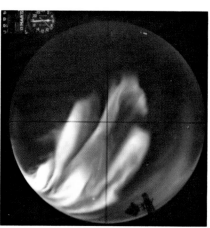

A strip of pictures taken with a modern all-sky camera that uses a wide-angle, fish-eye lens to image the upper hemisphere of the sky. These three exposures, taken within 40 seconds, indicate how rapidly the aurora may change.

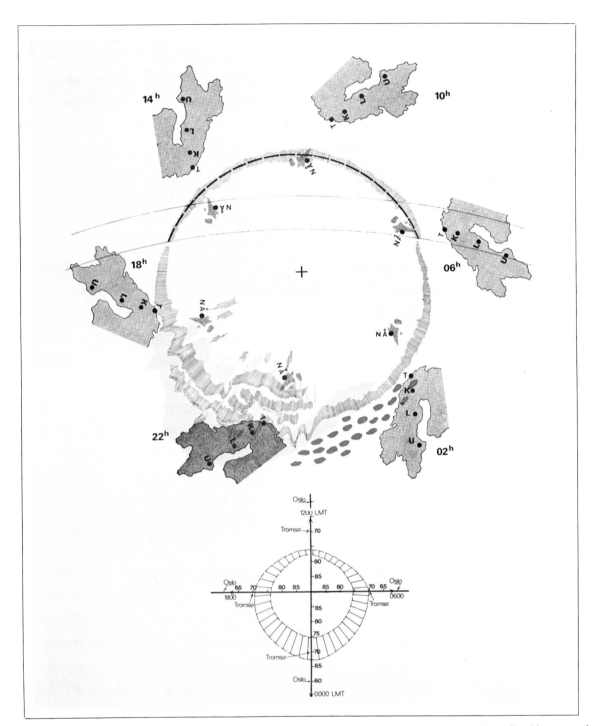

This diagram illustrates the concept of the auroral oval. The region of auroral occurrence is more or less fixed in space above the earth but centered on the geomagnetic rather than the geographic pole. As the earth rotates under the aurora the relative location of a country with respect to the auroral oval changes throughout the day. This diagram shows that Tromsø (T) in northern Norway is under the aurora between about 7 P.M. and 3 A.M. but is well south of the auoral region during the rest of the day. On the other hand, Nå in Spitzbergen is only under the aurora near 12 noon and is north of the auroral region the rest of the day and night. Note also that the character of the aurora varies with local time. In the evening hours well-defined arcs and bands are common, usually green in color; after midnight in the early morning hours the aurora is often patchy, and during the daytime hours (when aurora can be observed at these high latitudes only in the middle of winter) the aurora is less bright and usually red in color. Thus people in Nå not only see the aurora at a different time of day than do people in Tromsø, but they also see a different type of aurora.

(Left) A more compact version of the all-sky camera. Here the convex mirror reflects an image of the sky into the flat mirror held above. This mirror in turn is photographed by the camera which is located underneath the convex mirror and viewed through the hole in the mirror's center. A Plexiglass dome protects the all-sky camera from snow, etc. The installation was made at Fort Yukon, Alaska, in January 1972.

Photograph of the sun on December 21, 1957; this period was the peak of the sunspot cycle, and the solar disc shows numerous large sunspot groups. The resulting magnetic activity and auroras at the earth were recorded by the IGY network of observatories.

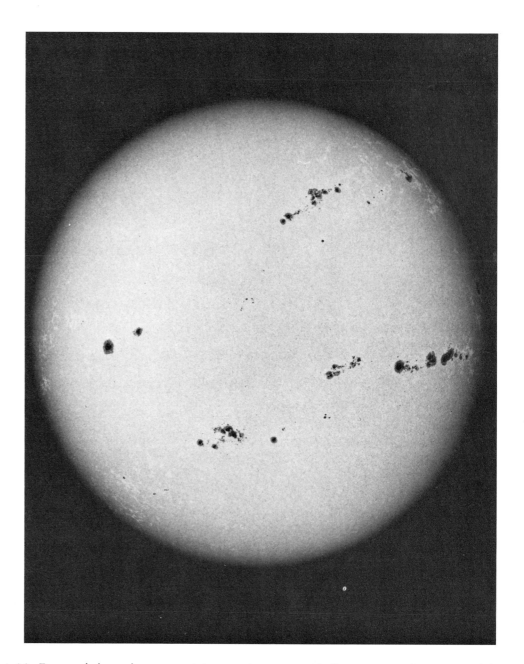

Air Force delayed any positive action, especially as no military need for satellites was envisaged. Satellites were neglected until 1954, when the National Science Foundation support for a small satellite launch in the forthcoming International Geophysical Year (1957–1958) led to a July 1955 White House announcement (163) that the United States would launch "small unmanned earth-circling satellites as part of the U.S. participation in IGY." A similar announcement 2 days later by the USSR (465) was more or less overlooked by the American press and public.

Heated interservice rivalry delayed the U.S. satellite program, and the rude awakening provided by *Sputnik I* in October 1957 should not have been unexpected. The absence of advance warning, in humiliating contrast to elaborate publicity about American progress towards a satellite launch, made the pill even more difficult to swallow in America. The second *Sputnik* launch, in November 1957, finally stirred activity in Washington, and the Army was authorized to activate a satellite program (the *Explorer* program) in conjunction with the existing Navy effort (the *Vanguard* program). An attempted Navy launch in December 1957 failed, and the first U.S. satellite, *Explorer I*, was launched by von Braun's Army group on January 21, 1958.

178

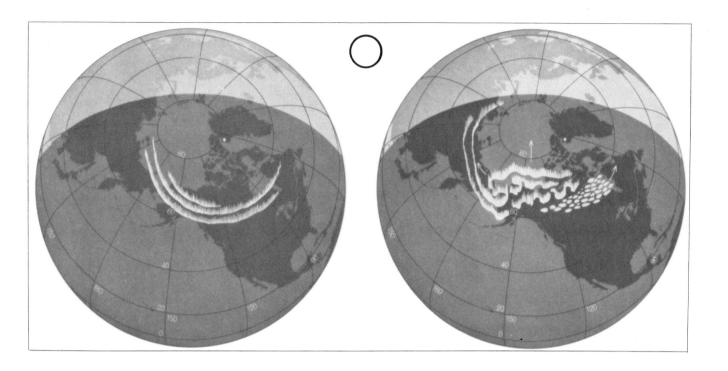

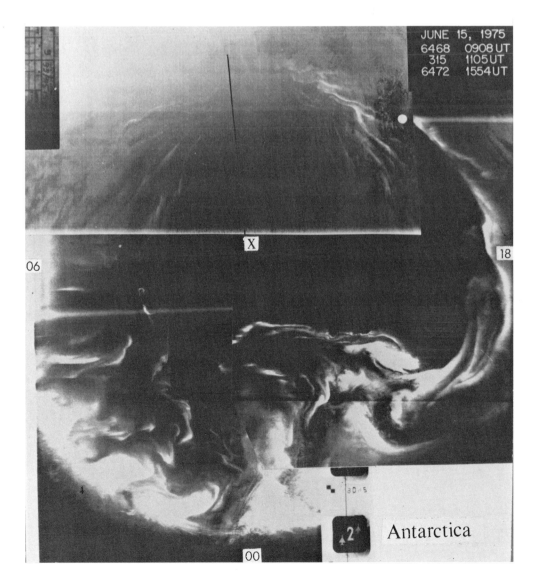

JUNE 15, 1975
6468   0908 UT
315   1105 UT
6472   1554 UT

06

X

18

00

2

Antarctica

The diagram on the top shows Akasofu's (1965) concept of the substorm, as derived from an analysis of all-sky camera photographs from the IGY. Data from many stations had to be used, as the field of view from a single station on the scale of this map is very small (indicated by the circle between the two maps).

The top left diagram shows quiet steady auroral arcs located along the nighttime part of the auroral oval. The top right diagram shows the dramatic brightening and spreading during a substorm. The bottom figure is a montage of three photographs taken by a satellite over Antarctica in 1975. (The solid white dot is the geographic south pole and the X indicates the geomagnetic pole). The whole auroral oval may be seen. A substorm is in progress and the form and extent of the bright substorm aurora agrees very well with Akasofu's picture derived some 10 years earlier.

TWENTY-FIVE CENTS

MAY 4, 1959

SPACE and the RADIATION BELT

TIME

E WEEKLY NEWSMAGAZINE

PHYSICIST
JAMES VAN ALLEN

$7.00 A YEAR

(REG. U.S. PAT. OFF.)

VOL. LXXIII NO. 18

The discovery of the radiation belt by James Van Allen
and his group in 1958–1959 received national attention.

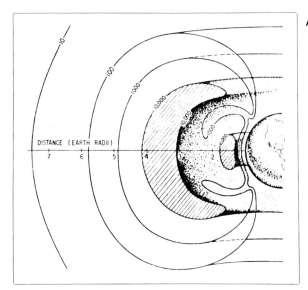

An early artist's concept of the Van Allen belts. The numbers on the contours indicate count rates of the geiger tube.

Carl McIlwain (left), James Van Allen (center), and George Ludwig (right) at the University of Iowa in 1958, with one of the earliest American satellite instruments.

Explorer I carried a small geiger counter supplied by James Van Allen's group at the University of Iowa. The instrument recorded the expected cosmic rays (very fast particles from deep space that enter the earth's magnetosphere) but unexpectedly showed no response when the satellite was at high altitudes over South America. There seemed no logical explanation, but the effect was confirmed by *Explorer III* 2 months later. (*Explorer II* went into the ocean.) Carl McIlwain of the Iowa group solved the problem, suggesting that the satellite entered a region of very intense energetic particles, so intense that they saturated the geiger tube and caused the counting circuits to read zero.

These results were first presented (480) by Van Allen and assistants on May 1, 1958, at a scientific meeting in Washington. They reported that

> . . . we believe that the extremely low output of the scaler is caused by very intense radiation which 'jams' the geiger tube so that it puts out pulses of such small height that they are below the threshold of the counting circuits. Laboratory tests show that this first happens for the present equipment when the radiation reaches . . . 35,000 counts per second.

The Eve of the Space Age

The following is excerpted from Walter Sullivan's book (450) on the IGY, *Assault on the Unknown*, published in 1961.

> It was the evening of October 4, 1957, and the ballroom of the Soviet Embassy in Washington was crowded with Soviet, American and other specialists in rocket research. As they stood, glasses of vodka in hand, the sound of their mingled voices filled the ornate room. The Russians were playing host to those attending a conference in Washington to coordinate plans for the launching of rockets and earth satellites during the International Geophysical Year.
>
> More than three months of the IGY already had passed, and much publicity had been given to the United States plans for launching a satellite, but the first attempt was not expected for at least two more months.
>
> At the conference, whose working sessions had come to an end that first day, the Russians had said they would make no advance announcement of their first attempt. They told their American colleagues, almost in so many words, 'We will not cackle until we have laid our egg.'
>
> As I stood in the ballroom with the scientists and a scattering of fellow newsmen, an official of the embassy said I was wanted on the telephone. It was *The New York Times'* Washington Bureau.
>
> 'Radio Moscow has just announced that the Russians have placed a satellite in orbit 900 kilometers above the earth,' the news editor said. Heart pounding with excitement, I bounded up the great staircase to the ballroom and threaded my way across the floor to pass the news on to Lloyd V. Berkner, the American member of the international committee running the IGY. Whatever the Russians there may have known of their government's intentions, they could not have known that the launching was successful. The Americans could at least have the pleasure of making the announcement to the Russians in their own embassy.
>
> Berkner clapped his hands loudly for silence and told the assembly of the launching . . . For the Russian delegates, the final day of the Washington conference was one of pent-up excitement released . . . In the Academy library, the model of the American Vanguard buzzed bravely, but a bit forlornly, around the globe.

The first American launch attempt two months later failed miserably when the first stage rocket exploded on the launch pad.

Front page of the *New York Times* for October 5, 1957, and February 1, 1958.

Launch of *Explorer I*, the first American satellite.

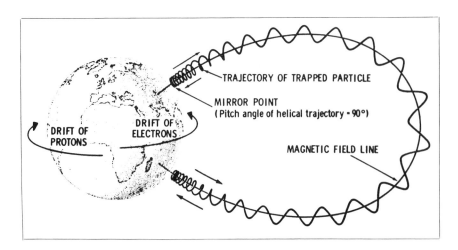

This diagram illustrates how electrically charged particles (protons and electrons) are trapped in the earth's magnetic field. The particles follow spiral trajectories about the magnetic field line and are reflected at low altitudes at the mirror point in each hemisphere. At the same time the particles slowly drift eastward (electrons) and westward (protons) around the earth. Some particles can continue these motions for days, months, even years, and it is the vast numbers of such particles around the earth that form the Van Allen belts.

The Van Allen radiation belts had been discovered!

The report that there was such intense radiation in space caused a sensation. It was "hot" enough to present a hazard to space travel. The Iowa group showed that the U.S. Department of Health danger level of 1.3 roentgens of radiation in 1 week would be received in less than 5 hours at the top of the *Explorer* orbit.

The idea that magnetic fields could confine large fluxes of energetic electrical particles was not new. Störmer had shown the theoretical possibility 50 years earlier (444) (see page 135). In 1956, Singer had postulated the existence of trapped particles in the outer regions of the earth's field (427). In fact, plans were already underway to test the trapping ability of the earth's field by injecting particles from a high-altitude nuclear explosion (450) (the Argus tests). But no one had suspected that the earth's field might contain such large fluxes of naturally occurring charged particles.

Russia belatedly claimed credit for the discovery but has never documented the claim. *Sputnik II* did carry a geiger tube but was at low altitudes over Russia. It was at the high altitudes of the radiation belt over Sydney, Australia, where scientists recorded the telemetry signals. But the Russians would not send the telemetry codes, so the Australians would not send the data (428).

There were all kinds of suggestions about sources of the trapped radiation. Both the Russians and the Americans accused each other of making the belt as a result of nuclear tests. No mechanism was suggested for getting the particles through the atmosphere to the height of the belts, so the accusations probably had political rather than scientific origin. False assumptions as to the types of particles composing the radiation belts led to suggestions that these particles could cause the aurora when they leaked into the atmosphere.

The IGY marked the emergence of science as a potent force in international affairs, a culmination of various international scientific efforts that had been gathering momentum for over a century. It had opened doors to the solar system, so to speak, and was the birth of magnetospheric physics. The Van Allen discovery was just the tip of the iceberg, and hundreds of increasingly sophisticated satellites have since been launched to explore the particle environment around our earth. Many such experiments were designed specifically to try to relate the rapidly developing new picture of the magnetosphere to the very old problem of the aurora.

Auroral physics had entered the space age!

14
The Aurora in Poetry and Literature

And now the Northern Lights begin to burn, faintly
at first, like sunbeams playing in the waters of the
blue sea. Then a soft crimson glow tinges the heavens.
There is a blush on the cheek of night. The colours
come and go; and change from crimson to gold, from gold
to crimson. The snow is stained with rosy light. Two-
fold from the zenith, east and west, flames a fiery
sword; and a broad band passes athwart the heavens, like
a summer sunset. Soft purple clouds come sailing over
the sky, and through their vapoury folds the winking
stars shine white as silver. With such pomp as this is
Merry Christmas ushered in, though only a single star
heralded the first Christmas.

Henry Wadsworth Longfellow
Driftwood — Frithiof's Saga, 1837

They rippled green with a wondrous sheen, they
fluttered out like a fan;
They spread with a blaze of rose-pink rays never
yet seen of man.
They writhed like a brood of angry snakes, hissing
and sulphur pale;
Then swift they changed to a dragon vast, lashing
a cloven tale.

Robert Service
The Ballad of the Northern Lights, 1909

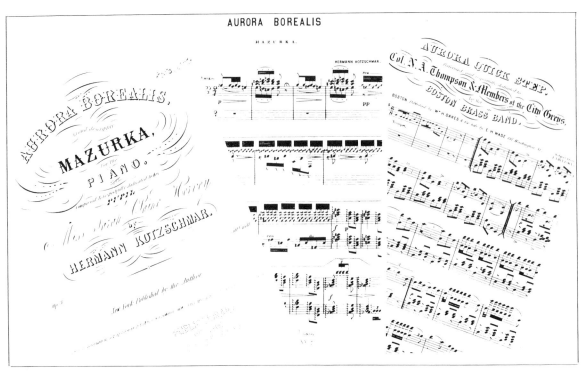

The aurora has inspired various musical scores.

Henry Wadsworth Longfellow
1807–1882
American poet and author

The earliest poetic references to the aurora are to be found in the epic poems of classical Rome. In reviewing these works, it is important to remember that the word "meteor" had a more general meaning than it does today and referred to any phenomenon or appearance in the heavens (see page 61). The word "comet" had the same meaning as it does in modern times, and it was widely believed that appearances of comets foretold the end of a monarch's reign.

There are two passages from Virgil (70–19 B.C.) that could refer to aurora (485), but such an inference is far from conclusive:

> . . . a sign from a cloudless sky . . . In the serene expanse of heaven they see arms, gleaming red in the clear air . . .
>
> *Aeneid,* Book 8, 524–529

> Never from a cloudless sky fell more lightnings; never so oft blazed fearful comets
>
> *Georgics,* Book 1, 487–488

Silius Italicus (26–101 A.D.) provides us with two passages (424) that can more definitely be interpreted as describing the aurora. In his *Punica,* concerning Hannibal and the Second Punic War, when the fighting was not going well against Marrus during the siege of Saguntum,

> Hannibal . . . flew off with frantic haste . . . the plume that nodded on his head showed a deadly brightness, even as a comet terrifies fierce kings, with its flaming tail and showers blood-red fire: the boding meteor sprouts forth ruddy rays from heaven, and the star flashes with a dreadful menacing light, threatening earth with destruction.
>
> *Punica,* Book 8, 460–467

Before the battle of Cannae, evil omens alarmed the soldiers:

> Javelins blazed up suddenly in the hands of astounded soldiers . . . seamen were frightened by fires burning on the mountains of Elba . . . and the bright hair of more than one comet, the portent that dethrones monarchs, showed its baleful glare.
>
> *Punica,* Book 8, 626–637

The clearest description of aurora by the classic poets of Rome is found in Marcus Lucanas' (39–65 A.D.) *The Civil War,* an epic poem (299) of 10 books describing the contest between Caesar and the Senate. In describing the fear of the citizens of Rome as they fled at the threat of an attack by Caesar, Lucanas wrote,

> . . . clear proof was given of worse to come, and the menacing gods filled earth, sky, and sea with portents. The darkness of night saw stars before unknown, the sky blazing with fire, lights shooting athwart the void of heaven, and the hair of the baleful star—the comet which portends change to monarchs. The lightning flashed incessantly in a sky of delusive clearness, and the fire, flickering in the heavens, took various shapes in the thick atmosphere, now flaming far like a javelin, and now like a torch with a fanlike tail. A thunderbolt, without noise of any clouds, gathered fire from the North and smote the capital of Latium.

The Civil War, Book 1, 522–533

Literary and poetic references to the aurora then seem rare for more than 1,000 years; in the thirteenth century, an English priest, Layamon, chronicled the life of the legendary King Arthur. In this chronicle (487) he has Arthur describe a dream, which could only have been inspired by a beautiful aurora:

> . . . Where I lay in slumber, and I gan for to sleep, methought that in the welkin came a marvelous beast, eastward in the sky, and loathsome in the sight; with lightning and with storm sternly he advanced; there is in no land any bear so loathly. Then came there westward, winding with the clouds, a burning dragon; burghs he swallowed; with his fire he lighted all this land's realm; methought in my sight that the sea gan to burn of light and of fire, that the dragon carried. This dragon and the bear, both together quickly soon together they came; they smote them together with fierce assaults; flames flew from their eyes as firebrands! Oft was the dragon above, and eftsoons beneath; nevertheless at the end high he gan rise, and he flew down right with fierce assault, and the bear he smote, so that he fell to the earth; and he there the bear slew, and limbmeal him tore. Then the fight was done, the dragon back went. This dream I dreamt, where I lay and slept.

It has been claimed (368) that the first poetic reference to the aurora after the Roman classicists is in Chaucer's *The Knight's Tale;* if so, it may also be the first use of the term "northern light," but the mention more probably describes light entering a north-facing door:

Geoffrey Chaucer 1340?–1400
English poet

> And ther-out cam a rage and such a vese
> That it made al the gates for to rese
> The northren light in at the dores shoon,
> For windowe on the wal ne was ther noon
> Thurgh which men mighten any light discerne.

There were at least two good auroral displays over England in Shakespeare's time (285) (in 1574 and 1583), and perhaps one of them was seen from Stratford-on-Avon and later inspired what seems to be the only probable auroral reference in Shakespeare's works (422).

> Fierce fiery warriors fought upon the clouds
> In ranks and squadrons and right form of war
> That drizzled blood upon the Capitol.

Julius Caesar II, 2,19

James Thomson 1700–1748
Scottish poet

The Scottish poet James Thomson (1700–1748) wrote a long work on the seasons (466), in which the poem *Summer,* first published in 1727, describes the aurora. The passage was afterward (1730) transferred with variations to *Autumn,* from which the following quotation is taken:

187

Oft in this season, silent from the north
A blaze of meteors shoots, ensweeping first
The lower skies, then all at once converge
High to the crown of heaven, and all at once
Relapsing quick, as quickly reascend
And mix and thwart, extinguish and renew,
All ether coursing in a maze of light.

From look to look contagious through the crowd
The panic runs, and into wondrous shapes
The appearance throws: armies in meet array
Throng with aerial spears and steeds of fire,
Till, the long lines of full-extended war
In bleeding fight commixed, the sanguine flood
Rolls a broad slaughter o'er the plains of heaven.

As thus they scan the visionary scene,
On all sides swells the superstitious din
Incontinent, and busy frenzy talks
Of blood and battle, cities overturned
And late at night in swallowing earthquake sunk
Or painted hideous with ascending flame,

Of sallow famine, inundation, storm,
Of pestilence and every great distress,
Empires subversed when ruling fate has struck
The unalterable hour: even Nature's self
Is deemed to totter on the brink of time.

Not so the man of philosophic eye
And inspect sage; the waving brightness he
Curious surveys, inquisitive yet unfixed
Of this appearance beautiful and new.

Presumably Thomson had witnessed the great aurora of 1716 that Halley described, as this was the only striking aurora to be seen in Scotland between Thomson's birth and the writing of his poem. This reappearance of aurora in England is also referred to by the poet William Collins in his ode, *Popular Superstitions of the Highlands of Scotland* (114).

As Boreas threw his young Aurora forth,
    In the first year of the first George's reign,
and battles rag'd in welkin of the North,
    They mourn'd in air, fell, fell Rebellion slain!

Richard Savage also mentions popular superstition (401) concerning the aurora in his poem *The Wanderer,* written in 1729.

See! . . . from the North, what streaming Meteors pour!
Beneath *Bootes* spring the radiant Train
And quiver thro' the Axle of his Wain.
O're Alters thus, impainted, we behold
Half-circling Glories shoot in Rays of Gold.
Cross *Ether* swift elance the vivid Fires!
And swift again each pointed Flame retires!
In *Fancy's* Eye encount'ring Armies glare,
And sanguine Ensigns wave unfurl'd in Air!
Hence the weak Vulgar deem impending Fate,
A Monarch ruined, or unpeopled State.

The Wanderer, Canto 3

188

In 1763, Christopher Smart wrote *A Song to David* (432), where the following lines appear.

Glorious the northern lights astream
Glorious the song, when God's the theme.

It may seem incongruous to us today that some scientific works have been written in poetic form, for example, the exposition of Democritus' atomic theory by Lucretius. In 1743 the Russian scientist Lomonosov wrote an ode to the northern lights (287):

But, where, O Nature, is thy law?
From the midnight lands comes up the dawn!
Is it not the sun setting his throne?
Is it not the icy seas that are flashing fire?
Lo, a cold flame has covered us!
Lo, in the night-time day has come upon the earth.
What makes a clear ray tremble in the night?
What strikes a slender flame in to the firmament?
Like lightning without storm clouds,
Climbs to the heights from earth?
How can it be that frozen steam
Should midst winter bring forth fire?

Rhetoric

Even more ambitious was a 157-page poem (in Latin) entitled *Aurora Boreali* written by the Jesuit Caroli Noceti in Rome in 1747 (350). The poem deals with history, mythology, and the current scientific ideas of the Russian academicians and of Mairan. Noceti comes out in favor of Mairan's explanation of auroras (see page 55). The text is illustrated with a complex geometrical figure of the earth and auroral regions and accompanied by 38 pages of detailed explanatory notes. A second edition in 1755 includes an Italian translation, but the work is apparently not available in English.

Christopher Smart 1722-1771
English poet

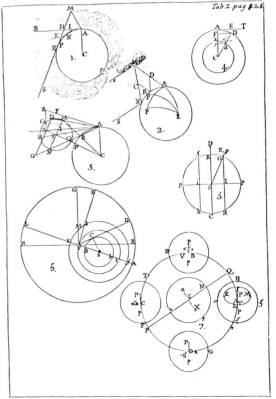

Title page and diagram for Noceti's
157 page poem on the aurora
(published in 1747).

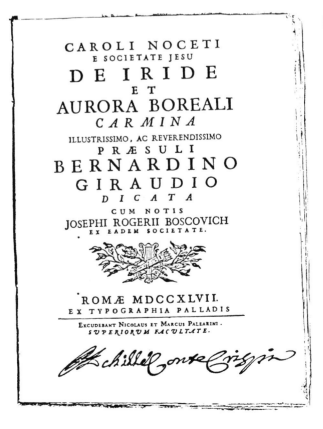

Samuel Taylor Coleridge 1772-1834
English poet and philosopher

Henry Hart Milman 1791-1868
English historian

Johann Wolfgang Goethe 1749-1832
German author, poet, and natural philosopher

Coleridge was a poet who expressed an eager interest in physical phenomena and avidly read Samuel Hearne's book (218) of his travels in arctic Canada. His references to aurora (113) range from the conventional, "streamy banners of the North," to the more inspired lines of the *The Rime of the Ancient Mariner*. When the long becalming finally ends,

> The upper air broke into life
> And a hundred fire-flags sheen
> To and fro they were hurried about
> And to and fro and in and out
> The wan stars danced between

The Rime of the Ancient Mariner

As a young man in the early 1800's, the historian H. H. Milman described an aurora borealis that he had seen (330). In his epic, *History of the Jews* (331), he comments that his lines "might seem to be, but were not, taken from the book of *Maccabees*," (see page 38):

> 'Twas midnight, but a rich untimely dawn
> Sheets the fir'd Arctic heaven; forth springs an arch,
> O'erspanning with a crystal pathway pure
> The starry sky: as though for Gods to march,
> With show of heavenly warfare daunting earth,
> To that wild revel of the northern clouds:
> They now with broad and bannery light distinct
> Stream in their restless waverings to and fro,
> And clash and cross, with hurtle and with flash
> Tilting their airy tournament . . .
> Never a brighter conflict in the skies
> Taught me that war was dear in Heaven: dream ye
> Of tamer faith in gentle Southern skies
> Your smooth and basking deities! our North
> Wooes not with tender hues and sunny smiles
> Soft worship, but emblazons all the air
> With semblance of celestial strife, unveils
> To us of their empyreal halls the pomp,
> The secret majesty of godlike war.

Samor, III

Aurora borealis observed on Col du Cabre in the French Alps at 1 A.M. in the morning (1859).

Pl. VI.

The following 10 plates are reproductions of paintings of sizes approximately 24″ x 36″, painted by the Danish artist Harald Moltke between 1899 and 1901. The eleventh plate appears on page 28. Moltke took part in two auroral expeditions from the Danish Meteorologisk Institut to Akureyri in Iceland in 1899-1900 and to Utsjoki in Finland in 1900-1901.

Pl. IV.

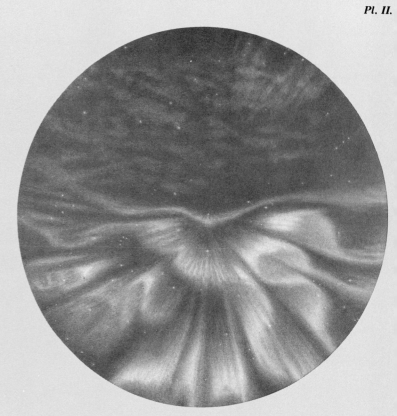

Pl. II.

Pl. XI.

Pl. IX.

Pl. I.

Pl. VIII.

Pl. X.

Pl. V.

193

Pl. VII.

A less metaphysical description of the aurora was penned by the American poet Carlos Wilcox (1794–1827) (502):

At length in northern skies, at first but small,
A sheet of light meteorous begun
To spread on either hand, and rise and fall
In waves, that slowly first, then quickly run
Along its edge, set thick but one by one
With spiry beams, that all at once shot high,
Like those through vapours from the setting sun;
Then sidelong as before the wind they fly,
Like streaking rain from clouds that flit along the sky.
                    Sights and Sounds of the Night

Goethe also made brief references to aurora; in a poem (198), *An Lida*, he refers to the eternal stars shimmering through the moving rays of the northern lights. Elsewhere Goethe wrote:

We come
from the far Pole and cold night
and would have liked to bring to your honour
the most beautiful northern lights.

Many other poets have briefly referred to the aurora in their poetry; and to try to compile an exhaustive list would be a Herculean task. From some better known English poets we have

The world would gaze upon those northern lights
Which flashed as far as where the musk-bull browses
                    Lord Byron
                    Don Juan, Canto 12

O'er his loins his trapping glow
Like the northern lights on snow
                    John Keats
                    To _____
                    (A Valentine to Georgiana Wylie)

But round the North, a light,
A belt, it seem'd of luminous vapor lay
                    Lord Tennyson
                    Sea Dreams

And red and bright in the streamers light
Were dancing in the glowing north.
                    Sir Walter Scott
                    The Lay of the Last Minstrel

And from north of the border, Robbie Burns wrote (84) of the elusiveness of the aurora:

Like the borealis race
That flit ere you can point the place
                    Robert Burns
                    Tam o'Shanter

Robert Southey in his *Curse of Kehama* writes (437),

Gleams of glory, streaks of flaming light
Openings of heaven, and streams that flash at night
In fitful splendour through the northern sky.

Lord Byron 1788–1824
English poet

John Keats 1795–1821
English poet

Alfred Lord Tennyson 1809–1892
English poet laureate

Sir Walter Scott 1771–1832
Scottish poet and novelist

Robert Burns   1759-1796
Scottish poet

Robert Southey   1774-1843
English poet laureate

One might expect frequent references in Russian literature, but the few that I know of are brief and matter-of-fact. The romantic poet Pushkin wrote in his novel in verse, *Eugene Onegin* (379),

> Tatyana . . . her rosy cheeks are not less bright
> than in the north aurora's blushes.

Both Dostoyevsky and Solzhenitsyn spent time in Siberia and traveled north of the arctic circle, yet the only mention of the aurora in the works of these authors seems to be in Solzhenitsyn's *Matryana's House* (436), in which Matryana sits,

> . . . her thoughts growing in wild profusion as they strove to catch the northern lights. As the light was fading . . . the owner's face looked yellow and ill, her bleary eyes showed how much illness had exhausted her.

One of the most beautiful poetic descriptions of the aurora is to be found in *Christmas Eve and Easter Day* by Robert Browning (81). Although

Painting of the aurora by Frank William Stokes. He was on the Peary Relief Expedition (1892) and later built an artist's studio at Inglefield Gulf (77.7°N) and spent the 1893-1894 winter there to paint views of the Arctic. Stokes employed considerable artistic license, so much so that the leader of the expedition published a disclaimer of any agreement of the painting with reality.

various literary criticisms refer to the relevant passages as a report of a vision or a dream or a product of the imagination, there can be no doubt that Browning's words were inspired by an aurora that he had seen some years previously. The poem was written in Italy in 1850 and refers to an incident on Easter Eve 3 years before. The year 1847 was at the peak of the sunspot cycle, and many beautiful auroras were seen throughout Europe that year. However, Easter Eve fell on April 3 in 1847, and auroral catalogs do not report unusual activity on that night. (There were great auroras on March 19 and April 7.) This apparent contradiction seems to argue against an auroral inspiration for the poetry, but there are other reasons why Browning could have wished the incident to be on Easter Eve, so perhaps we can excuse the disagreement in exact dates as poetic license.

Easter Day is not regarded as one of Browning's best poems, but it is widely studied for its revelation of the poet's religious beliefs. Certainly, the passages describing his "vision," or "dream," stand out from the body of the poem.

Robert Browning 1812-1889
English poet

xv
. . . throwing back my head
With light complacent laugh, I found
Suddenly all the midnight round
One fire. The dome of heaven had stood
As made up of a multitude
Of handbreadth cloudlets, one vast rack
Of ripples infinite and black,
From sky to sky. Sudden there went,
Like horror and astonishment,
A fierce vindictive scribble of red
Quick flame across, as if one said
(The angry scribe of Judgment) "There—
"Burn it!" And straight I was aware
That the whole ribwork round, minute
Cloud touching cloud beyond compute,
Was tinted, each with its own spot
Of burning at the core, till clot
Jammed against clot, and spilt its fire
Over heaven, which 'gan suspire
As fanned to measure equable,—
Just so great conflagrations kill
Night overhead, and rise and sink,
Reflected. Now the fire would shrink
And wither off the blasted face
Of heaven, and I distinct might trace
The sharp black ridgy outlines left
Unburned like network—then, each cleft
The fire had been sucked back into,
Regorged, and out it surging flew
Furiously, and night writhed inflamed,
Till, tolerating to be tamed
No longer, certain rays world-wide
Shot downwardly. On every side
Caught past escape, the earth was lit,
As if a dragon's nostril split
And all his famished ire o'erflowed;
Then, as he winced at his lord's goad
Back he inhaled: whereat I found
The clouds into vast pillars bound,
Based on the corners of the earth,
Propping the skies at top: a dearth
Of fire i' the violet intervals,
Leaving exposed the utmost walls
Of time, about to tumble in
And end the world.

xvi
I felt begin
the Judgment Day: . . .

xvii
The final belch of fire like blood,
Overbroke all heaven in one flood
Of doom. Then fire was sky, and sky
Fire, and both, one brief ecstasy,
Then ashes. But I heard no noise
(Whatever was) because a voice
Beside me spoke thus, "Life is done,
Time ends, Eternity's begun,
And thou are judged for evermore."

xviii
I looked up: all seemed as before;
Of that cloud Tophet overhead
No trace was left: I saw instead
The common round me, and the sky
Above, stretched drear and emptily
Of life . . .

"A dream—a waking dream at most"
(I spoke out quick, that I might shake
The horrid nightmare off, and wake.) . . .

xxxiii
When I lived again
The day was breaking—the grey plain
I rose from, silvered thick with dew
Was this a vision? False or true?
Since then, three varied years are spent,
And commonly my mind is bent
To think it was a dream—be sure
A mere dream and distemperature—
The last day's watching: then the night.—
The shock of that strange Northern Light
Set my head swimming, bred in me
A dream . . .

Robert Browning
Easter Day

197

Christopher Pearse Cranch 1813–1892
American painter and poet

The American painter and poet Christopher Cranch wrote *To The Aurora Borealis* during the year he spent as a preacher in south Boston (1840) (123).

Arctic fount of holiest light
Springing through the winter night,
Spreading far beyond yon hill
When the earth is dark and still,
Rippling o'er the stars, as streams
Ripple o'er their pebble-gleams—
Oh, for names, thou vision fair,
To express thy splendors rare!
. . . .

Is not human fantasy,
Wild Aurora, likest thee,
Blossoming in nightly dreams
Like thy shifting meteor-gleams?
. . . .

Reaching upwards from the earth
To the *Soul* that gave it birth.
When the noiseless beck of night
summons out the *inner* light . . .

Many poems about the aurora appear in the various publications of wintering expeditions in the Arctic. Though not written by "professional" poets, these poems succeed in conveying the mystery and inspiration of the aurora. Here we quote some lines from *The North Georgia Gazette and Winter Chronicle*, published by Parry's expedition (397, 489).

Transfix'd with wonder on the frozen flood,
The blaze of grandeur fired my youthful blood;

Deep in th' o'erwhelming maze of Nature's laws,
'Midst her mysterious gloom, I sought the cause;
But vain the search!

> James Clure Ross
> *Lines Suggested by the Brilliant Aurora,
> Jan. 15, 1820.*

Beyond yon cloud a stream of paly light
Shoots up its pointed spires—again immerged,
Sweeps forth with sudden start, and waving round
In changeful forms, assumes the brighter glow
Of orient topaz—then as sudden sinks
In deeper russet, and at once expires!

> Cyrus Wakeham
> *Reflections on the Morning of Christmas
> Day, 1819, North Georgia*

Many poets have written poems with the title *Northern Lights*, and a few selected quotations follow:

Was ever such a vision to mortals sent
As Northern Lights in the heavens flaming?

> Einar Benediktsson
> *Northern Lights*

Benjamin Franklin Taylor's poem (462) gives an amusing explanation of the auroral phenomenon:

> To claim the Arctic came the Sun,
> With banners of the burning zone;
> Unrolled upon their airy spars
> They froze beneath the light of stars;
> And there they float, those streamers old,
> Those Northern Lights, forever cold!

Benjamin Franklin Taylor
The Northern Lights

Benjamin Franklin Taylor 1819-1887
American poet

The awe and uneasiness with which many have viewed the aurora in Scotland is expressed in the following two quotations:

> "Ma daddy turns tae the sky
>   And cries on me tae see
> They shiftin beams that dance oot-by
>   And fleg the he'rt o' me."
> "Laddie, the North is a' a-lowe
>   Wi' fires o' siller green,
> The stars are dairk owre Windyknowe
>   That were sae bricht the streen,
>
> "The lift is fu' o' wings o' licht
>   Risin' an' deein' doon —"
> "Rax ye yer airm and haud it ticht
>   Aboot yer little loon,
> For oh! the North's an eerie land
>   And eerie voices blaw
> Frae whaur the ghaists o' deid men stand
>   Wi' their feet amangst the snaw;
> "And owre their heids the midnicht sun
>   Hangs like a croon o' flame,
> It's i' the North you licht's begun
>   An' I'm fear'd that it's the same!
>
> Haud ye me ticht! Oh, div ye ken
>   Gid sic-like things can be
> That's past the sicht o' muckle men
>   And nane but bairns can see?"

Violet Jacob
The Northern Lights

W. E. Aytoun 1813-1865
Scottish poet

> All night long the northern streamers shot
>   across the trembling sky;
> Fearful lights that never beckon;
> Save when kings and heroes die.

W. E. Aytoun
Edinburg after Flodden

John Greenleaf Whittier   1807-1892
American poet

John Greenleaf Whittier also wrote of the aurora (499) as an omen:

A light is troubling Heaven! — A strange, dull glow
Is trembling like a fiery veil between
The blue sky and the earth; . . .
. . . Wherefore then
Burns the strange fire in Heaven? — It is as if
Nature's last curse — . . .

        Lo — a change!
The fiery flashes sink, and all along
The dim horizon of the fearful North
Rests a broad crimson, like a sea of blood,
Untroubled by a wave. And lo — above,
Beneath a luminous arch of pale, white,
Clearly contrasted with the blue above
And the dark red beneath it. Glorious!
How like a pathway for the sainted ones — . . .

Another change. Strange, fiery forms uprise
On the wide arch, and take the throngful shape
Of warriors gathering to the strife on high,
A dreadful marching of infernal shapes,
Beings of fire with plumes of bloody red,
With banners flapping o'er their crowded ranks,
And long swords quivering up against the sky!
And now they meet and mingle; and the ear
Listens with painful earnestness to catch
The ring of cloven helmets and the groan
Of the down-trodden. But there comes no sound . . .
It is as if the dead had risen up
To battle with each other — . . .

Steed, plume, and warrior vanish one by one,
Wavering and charging to unshapely flame; . . .

        It has passed —
And Heaven again is quiet; and its stars
Smile down serenely. . . .
. . . But the hearts
Of those who gazed upon it, yet retained
The shadow of its awe — the chilling fear
Of its ill-boding aspect. It is deemed
A revelation of the things to come —
Of war and its calamities — the storm
Of the pitched battle, and the midnight strife
Of heathen inroad — the devouring flame,
The dripping tomahawk, the naked knife. . . .
The Aërial Omens

The Canadian poet J. K. Foran wrote a poem, *The Aurora Borealis*, which beautifully conveys the dynamic nature of the aurora (172), slowly building and teasing with glimpses of what is to come, then a crescendo of action, ending with a gentle fading of activity:

Joseph Kearney Foran 1857-1931
Canadian poet

In the north, behold a flushing:
Then a deep and crimson blushing;
Follow'd by an airy rushing
Of the purple waves that rise:
As when armed host advances,
See, a silver banner dances,
And a thousand golden lances
Shimmer in the Boreal skies!
    The picture slowly dies!

Now, in bright prismatic splendor,
Comes a vision still more tender,
As a curtain white and slender
Falls across the space afar;
Where its lacy folds are ending,
With the black of distance blending,
Are its miles of fringe descending,
Hanging from a golden bar —
    Pinned to heav'n by a star!

Like a monster rous'd from sleeping,
First to westward slowly creeping,
Then, in headlong fury, sweeping,
Rush'd a mammoth cloud of black;
Rolling upward, plunging, lashing,
Through the fairy curtain dashing,
With a thousand beauties flashing,
O'er its phosphorescent back —
    Endless streamers in its track!

Visions of Arabian story:
Crimson fields of battle glory;
In kaleidoscopic glory,
Shifting, fading, restless tents;
Fairy armies wild in motion;
Jewell'd shrines of strange devotion;
And a greenish, tideless ocean,
Bound by ice-clad mounts and dents.
    Saw we through the curtain's rents!

Transformations still beholding,
Up the veil is swiftly folding —
And fantastic shapes are moulding
On the background of the sky;
Dimmer armies are parading, —
Fainter wreaths the light is brading,
While the splendors all are fading
Into the deep purple dye,
    Disappearing from the eye!

What a mighty revelation
In the wonders of creation!
Joy and grief and expectation
Dance through nature's scenes at night!
Life is dark'ning, life is glowing,
Pleasant zephyrs round it blowing,
Brilliant colors through it flowing,
Fading all when once in sight;
    Such is life — a Northern Light!

Emily Dickinson 1830–1886
American poet

Wallace Stevens 1879–1955
American poet

Emily Dickinson expresses the inspiration she received on seeing a fine aurora, coupled with feelings of personal insignificance that such a display evokes (141):

Of bronze and blaze
The north, to-night!
So adequate its forms,
So preconcerted with itself,
So distant to alarms —
An unconcern so sovereign
To universe, or me,
It paints my simple spirit
With tints of majesty,
Till I take vaster attitudes,
And strut upon my stem,
Disdaining men and oxygen,
For arrogance of them.
My splendours are menagerie;
But their competeless show
Will entertain the centuries
When I am, long ago,
An island in dishonoured grass,
Whom none but daisies know.
                    Aurora

In another poem, she expresses the sense of privilege she feels on seeing an aurora.

Morning is due to all—
To some—the night—
To an imperial few—
The Auroral Light.
                    Poem 1577

More recently, the American poet Wallace Stevens wrote a disturbing and complex poem (440) entitled *The Auroras of Autumn,* in which the aurora is used as the symbol of a paradox. The aurora borealis, seen as the serpent of appearance, is a withdrawal of the colors, deceptions, and dangers of life into a nest in the northern sky:

This is where the serpent lives, the bodiless.
His head is air. Beneath his tip at night
Eyes open and fix on us in every sky

[This is man's nest, his final place:]

Or is this another wriggling out of the egg
Another image at the end of a cave
Another bodiless for the bodies slough? . . .

[The northern lights streak red the midnight sky:]

These lights may finally attain a pole
In the midmost midnight and find the serpent there,

In another nest, the master of the maze
Of body and air and forms and images
Relentlessly in possession of happiness

[The ultimate poison is that man cannot believe its withdrawal can ever be so complete that it becomes its opposite—a polar brightness in the midst of midnight, a complete happiness beyond life in the midst of death:]

This is his poison; that we should disbelieve
Even that . . .

With its frigid brilliance, its blue-red sweeps
And gusts of great enkindlings, its polar green
The color of ice and fire and solitude . . .

It is a theatre floating through the clouds
Itself a cloud, although of misted rock
And mountains running like water, wave on wave,

Through waves of light. It is a cloud transformed
To cloud transformed again, idly the way
A season changes color to no end,

Except the lavishing of itself in change . . .

The cloud drifts idly through half-thought-of forms . . .

The scholar of one candle sees
An Arctic effulgence flaring on the frame
Of everything he is. And he feels afraid. . . .

It is a thing of ether that exists
Almost as predicate. But it exists,
It exists, it is visible, it is, it is.

So then, these lights are not a spell of light,
A saying out of cloud, but innocence.
An innocence of the earth, and no false sign

Or symbol of malice . . .

[Finally the lights shed an aurora of brilliance and hope:]

The stars are putting on their glittering belts.
They throw around their shoulders cloaks that flash . . .

Like a blaze of summer straw, in winter's nick.

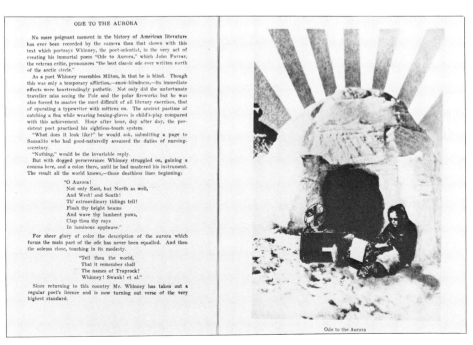

Pages from a humorous fictional book on arctic exploration by W. E. Traprock published in 1922.

203

Painting by Stephen Hamilton of the
aurora from Labrador.

The aurora has also given inspiration to the theatre. *The Twilight of the Gods* has a closing scene in which the heavens and earth turn red with blood (488); it is interesting to note that in the stage directions for this work, which were written just after a period of marked auroral activity in 1846, Wagner says the destruction of Valhalla should be indicated by "a red glow similar to an aurora."

An auroral reference is also found in one of the most talked of plays of early this century (456), John M. Synge's *Playboy of the Western World:*

And I'd be as happy as a lark on St. Patrick's Day
Watching the lights pass in the north
Smoking me pipe and drinking me fill.

There are, of course, many beautiful prose descriptions of particular auroras to be found in books scattered throughout almost every library classification, especially in published works arising from many expeditions of the past 100 years to explore the polar regions. Many have already been quoted in other chapters, and one could easily fill an entire book with such descriptions. Consequently, the following can only represent a small personal selection.

Weyprecht, who wintered in Franz-Joseph's Land with the Austrian-Hungarian Arctic Expedition of 1872–1873 concluded a beautiful description of auroras with the following words (495):

Karl Weyprecht 1838–1881
Austrian arctic explorer, a principal organizer of the First Polar Year

> From the centre issues a sea of flames: is that sea red, white, or green? Who can say? It is all three colours at the same moment. The rays reach almost to the horizon: the whole sky is in flames. Nature displays before us such an exhibition of fireworks as transcends the power of imagination to conceive. Involuntarily we listen: such a spectacle must, we think be accompanied with sound. But unbroken stillness prevails; not the least sound strikes the ear. Once more it becomes clear over the ice, and the whole phenomenon has disappeared with the same inconceivable rapidity with which it came, and gloomy night has again stretched her dark veil over everything. This was the Aurora of the coming storm — the Aurora in its fullest splendour. No pencil can draw it, no colours can paint it, and no words can describe it in all its magnificence.

The famous Norwegian arctic explorer Fridtjof Nansen was always fascinated by the aurora, and many beautiful descriptions may be found in the books of his arctic adventures. Perhaps Nansen's most impressive auroral prose (345) was inspired by a spectacular display he witnessed on the Fram Expedition while trapped through the long arctic winter in the frozen pack ice:

Fridtjof Nansen 1861–1930
Famed Norwegian arctic explorer, author, and statesman. Nobel Peace Prize winner in 1922. His books contain numerous descriptions of the northern lights.

> Later in the evening Hansen came down to give notice of what really was a remarkable appearance of aurora borealis. The deck was brightly illuminated by it, and reflections of its light played all over the ice. The whole sky was ablaze with it, but it was brightest in the south; high up in that direction glowed waving masses of fire. Later still Hansen came again to say that now it was quite extraordinary. No words can depict the glory that met our eyes. The glowing firemasses had divided into glistening, many-coloured bands, which were writhing and twisting across the sky both in the south and north. The rays sparkled with the purest, most crystalline rainbow colours, chiefly violet red or carmine and the clearest green. Most frequently the rays of the arch were red at the ends, and changed higher up into sparkling green, which quite at the top turned darker, and went over into blue or violet before disappearing in the blue of the sky; or the rays in one and the same arch might change from clear red to clear green, coming and going as if driven by a storm. It was an endless phantasmagoria of sparkling colour, surpassing anything that one can dream. Sometimes the spectacle reached such a climax that one's breath was taken away; one felt that now something extraordinary must happen — at the very least the sky must fall. But as one stands in breathless expectation, down the whole thing trips, as if in a few quick, light scale-runs, into bare nothingness. There is something most undramatic about such a *dénouement*, but it is all done with such confident assurance that one cannot take it amiss; one feels one's self in the presence of a master who has the complete command of his instrument. With a single stroke of the bow he descends lightly and elegantly from the height of passion into quiet, every-day strains, only with a few more strokes to work himself up into passion again. It seems as if he were trying to mock, to tease us. When we are on the point of going below, driven by 61 degrees of frost (–34°C), such magnificent tones again vibrate over the strings that we stay, until noses and ears are frozen. For a finale, there is a wild display of fireworks in every tint of flame — such a conflagration that one expects every minute to have it down on the ice, because there is not room for it in the sky. But I can hold out no longer. Thinly dressed, without a proper cap, and without gloves, I have no feeling left in body or limbs, and I crawl away below.

One of Nansen's woodcuts of the aurora.

Bayard Taylor 1825–1878
American poet and writer

Richard Evelyn Byrd 1888–1957
American aviator and explorer

In a book (461) that describes his travels through arctic Scandinavia in the 1850's, Bayard Taylor wrote an inspired description of the first aurora he witnessed:

I looked upward, and saw a narrow belt or scarf of silver fire stretching directly across the zenith, with its loose, frayed ends slowly swaying to and fro down the slopes of the sky. Presently it began to waver, bending back and forth, sometimes slowly, sometimes with quick, springing motion, as if testing its elasticity. Now it took the shape of a bow, now undulated into Hogarth's line of beauty, brightening and fading in its sinuous motion, and finally formed a shepherd's crook, the end of which suddenly began to separate and fall off, as if driven by a strong wind, until the whole belt shot away in long, drifting lines of fiery snow. It then gathered again into a dozen dancing fragments, which alternately advanced and retreated, shot hither and thither, against and across each other, blazed out in yellow and rosy gleams or paled again, playing a thousand fantastic pranks, as if guided by some wild whim.

We lay silent, with upturned faces, watching this wonderful spectacle. Suddenly, the scattered lights ran together, as by a common impulse, joined their bright ends, twisted them through each other, and fell in a broad, luminous curtain straight downward through the air until its fringed hem swung apparently but a few yards over our heads. This phenomenon was so unexpected and startling that for a moment I thought our faces would be touched by the skirts of the glorious auroral drapery. It did not follow the spheric curve of the firmament, but hung plumb from the zenith, falling, apparently, millions of leagues through the air, its folds gathered together among the stars and its embroidery of flame sweeping the earth and shedding a pale, unearthly radiance over the wastes of snow. A moment afterwards and it was again drawn up, parted, waved its flambeaux and shot its lances hither and thither, advancing and retreating as before. Anything so strange, so capricious, so wonderful, so gloriously beautiful, I scarcely hope to see again.

The American explorer, aviator, and scientist, Admiral Byrd, was also a very creative writer. In the 1930's he spent an antarctic winter completely alone at a small station on the antarctic continent. His reasons were ostensibly to get far enough south to study the aurora, but in his beautiful book *Alone* he talks of his passion to get away from civilization and all people. His was an epic of survival, and sometime during that winter he wrote this description (86) of the aurora australis:

The wind blew gently from the Pole, and the temperature was between 40 and 50 below. When Antarctica displays her beauty, she seems to give pause to the winds, which at such times are always still. Overhead, the Aurora began to change its shape and become a great, lustrous serpent moving slowly across the zenith. The small patch in the eastern sky now expanded and grew brighter; and almost at the same instant folds in the curtain over the Pole began to undulate, as if stirred by a celestial presence. Star after star disappeared as the serpentine folds covered them. It was like watching a tragedy on a cosmic scale; the serpent, representing the forces of evil, was annihilating the beauty.

Suddenly, the serpent disappeared. Where it had been only a moment before, the sky was clear — the stars looked as if they had never been dimmed. When I looked for the luminous patch in the eastern sky, it too, had disappeared. I was left with the tingling feeling that I had witnessed a sight denied to all other mortal men.

From another time and place the famous Jesuit writer and poet Gerard Manley Hopkins wrote in his diary (240) in 1870:

Sept. 24.   First saw the Northern Lights. My eye was caught by beams of light and dark very like the crown of horny rays the sun makes behind a cloud. At first I thought of silvery cloud until I saw that these were more luminous and did not dim the clearness of the stars in the Bear. They rose slightly radiating thrown out from the earthline. Then I saw soft pulses of light one after another rise and pass upwards arched in shape but waveringly and with the arch broken. They seemed to float, not following the warp of the sphere as falling stars look to do but free though concentrical with it. This busy working of nature wholly independent of the earth and seeming to go on in a strain of time not reckoned by our reckoning of days and years but simpler and as if correcting the preoccupation of the world by being preoccupied with and appealing to and dated to the day of judgement was like a new witness to God and filled me with delightful fear.

A description by T. C. Haliburton (368) in his story *Old Judge*:

The sun has scarcely set behind the dark wavy outline of the western hills, ere the aurora borealis mimics its setting beams, and revels with wild delight in the heavens, which it claims as its own, now ascending with meteor speed to the zenith, then dissolving into a thousand rays of variegated light, that vie with each other which shall first reach the horizon; now flashing bright, brilliant and glowing, as emanations of the sun, then slowly retreating from view pale and silvery white like wandering moonbeams.

The beauty of northern lights has thus inspired famous poets and writers throughout the centuries. In fact, one suspects that any writer fortunate enough to witness a fine aurora would feel impelled to describe that experience in his poems or stories. Even those who are not writers by profession are moved to write lyrically about the aurora; let me quote from a letter from Mrs. Alice Colp of Canton, Massachusetts. Mrs. Colp wrote to me recently that

. . . I hope that you might explain a beautiful sight that I saw in the sky about ten years ago. At that time I owned and worked in a small variety store here in Canton [Massachusetts] and worked long hours, so that I would be closing shop and walking home about 10 p.m. every night. This particular night was quite cold but very calm, and as I started up the street, I noticed the long streaks of light popping up the sky. I had seen "northern lights" before, so paid no attention to them. My home at that time was about a five minute walk from the store beside Bolivar Pond. The night was calm and peaceful, I stopped for a minute by the pond and looked up at the sky. I shall never forget the beautiful sight I saw. I can only describe it as being inside a beautiful church, with intricate columns raising to the center in perfect design. As I turned around, it was all round me, and I was in the center. For the next couple of days, I watched the newspapers to see if anyone had reported this beautiful sight. Everyone I told about it must have thought it was my imagination.

Gerald Manley Hopkins 1844-1889
English poet and Jesuit

AURORA. GUILDFORD. OCT. 24. 1870.
FROM A WATER COLOUR DRAWING.

This water color by Capron of an aurora seen from Guilford, England (October 24, 1870) beautifully illustrates the description by Mrs. Colp (about 1973).

207

Tigermine

Conrad Aiken 1889-1973
American poet and novelist

In your lonely luminous palace
  of Aurora Borealis
lovely lonely Tigermine
  burn and shine flame and shine.
Smoothly purr your snow-white purr
  blue-eyed lick your ermine fur
smile that smile that at a mile
  still looks only like a smile.
Clean with icicles your claws
  preen with snow your tufted paws
scour your eyes with frozen tears
  scoop the snow-pearls from your ears.
Wind your tail and wind it thrice
  three times round a block of ice
just to show that winds that blow
  and the arctic sleet and snow
are the things that you love best
  snowflake on a snowdrift's breast.
O how blue, how blue those eyes
  like blue stars in snow-white skies
O how bright, how bright they shine
  lonely lovely Tigermine.
Strange you never DO get thinner
  with one snowflake for your dinner
strange you never want to warm
  tail or paw from howling storm
strange you like to live up there
  in that O so Polar air!
But I'll tell you why it's so:
  Tigermine was MADE of snow
and he's happy as can be
  gazing on that frozen sea
purring loudly purring proudly
  gazing on that endless sea
in his lone electric palace
  of Aurora Borealis
in the flickering fleeting lights
  of his O so northern nights.

The children's poem *Tigermine*
by Conrad Aiken illustrated by
John Vernon Lord.

Aurora and Alliteration

In August–September 1859, the steamer *Europa* sailed from Boston to Liverpool. This was one solar rotation after the great aurora of 1859, and almost every evening there were beautiful auroral displays. An anonymous gentleman, H.C.B., and two fellow passengers "passed away an hour or two in attempting to compose a poem on the Aurora—following the alphabetical system (217). Composed hastily, and jotted down by the light of the binacle lamp . . . "

An Artful And Amusing Attempt At Alphabetical Alliteration Addressing Aurora

Awake Aurora! And Across All Airs
By Brilliant Blazon Banish Boreal Bears,
Crossing Cold Canope's Celestrial Crown.
Deep Darts Descending Dive Delusive Down.
Entranced Each Eve "Europa's" Every Eye
Firm Fixed Forever Fastens Faithfully,
Greets Golden Guerdon Gloriously Grand:
How Holy Heaven Holds High His Hollow Hand!
Ignoble Ignorance, Inapt Indeed—
Jeers Jestingly Just Jupiter's Jereed:
Knavish Khamschatkans, Knightly Kurdsman Know
Long Labrador's Light Lustre Looming Low;
Midst Myriad Multitudes Majestic Might
No Nature Nobler Numbers Neptune's Night.
Opal of Oxus Or Ophir's Ores
Pale Pyrrhic Pyres Prismatic Purple Pours—
Quiescent Quivering, Quickly, Quaintly Queer,
Rich, Rosy, Regal Rays Resplendent Rear;
Strange Shooting Streamers Streaking Starry Skies
Trial Their Triumphant Tresses—Trembling Ties.
Unseen, Unhonoured Ursa—Underneath
Veiled, Vanquished—Vainly Vying—Vanisheth:
Wild Woden, Warning, Watchful—Whispers Wan
Xanthitic Xeres, Xerxes, Xenophon,
Yet Yielding Yesternight Yules Yell Yawns
Zenith's Zebraic Zizzag, Zodiac Zones.

The Ballad of the Northern Lights

One of the best-loved balladeers of the English speaking world is Robert Service (416). His ballads ring with a virility and humor and a gaiety and pathos that cannot help but strike a responsive cord in all of us. Service spent much of his life in the Yukon and Alaska, and his most famous ballads are of the rough-and-tumble pioneer life of the north. He was a keen observer of nature, and his awe and appreciation of the aurora may be seen in many lines from his poetry.

Robert Service 1876-1958
Canadian "Poet of the Yukon"

There where the livid tundras keep their tryst with the tranquil snows;
There where the silences are spawned, and the light of hell-fire flows
Into the bowl of the midnight sky, violet, amber and rose.

The Heart of the Sourdough

In the hush of my mountained vastness, in the flush of my midnight skies.

The Law of the Yukon

This is the song of the parson's son, as he squats in his shack alone,
On the wild, weird nights, when the Northern Lights shoot up from the frozen zone,
And it's sixty below, and couched in the snow the hungry huskies moan:

The Parson's Son

Were you ever out in the Great Alone, when the moon was awful clear,
And the icy mountains hemmed you in with a silence you most could hear;
With only the howl of a timber wolf, and you camped there in the cold,
A half-dead thing in a stark, dead world, clean mad for the muck called gold;
While high overhead, green, yellow and red, the North Lights swept in bars?—
Then you've a hunch what the music meant . . . hunger and night and the stars.

The Shooting of Dan McGrew

There are strange things done in the midnight sun
　By the men who moil for gold;
The Arctic trails have their secret tales
　That would make your blood run cold;
The Northern Lights have seen queer sights,
　But the queerest they ever did see
Was that night on the marge of Lake Lebarge
　I cremated Sam McGee.

The Cremation of Sam McGee

This is the tale that was told to me by the man with the crystal eye,
As I smoked my pipe in the camp-fire light, and the Glories swept the sky;
As the Northlights gleamed and curved and streamed, and the bottle of
　"hooch" was dry.

. . . .

This was the tale he told to me, that man so warped and gray,
Ere he slept and dreamed, and the camp-fire gleamed in his eye in a
　wolfish way—
That crystal eye that raked the sky in the weird Auroral ray.

The Ballad of One-Eyed Mike

Am I too old, I wonder? Can I take one trip more?
Go to the granite-ribbed valleys, flooded with sunset wine,
Peaks that pierce the aurora, rivers I must explore,
Lakes of a thousand islands, millioning hordes of the Pine?

The Nostomaniac

In Rory Borealis Land the winter's long and black.
The silence seems a solid thing, shot through with wolfish woe;
And rowelled by the eager stars the skies vault vastly back.
And man seems but a little mite on that weird-lit plateau.
Nothing to do but smoke and yarn of wild and misspent lives,
Beside the camp-fire there we sat—what tales you told to me
Of love and hate, and chance and fate, and temporary wives!
In Rory Borealis Land, beside the Arctic Sea.

While the Bannock Bakes

Service also made the aurora the topic of a ballad of its own. In *The Ballad of the Northern Lights* he weaves in the beauty, the mystery, and the awe of the aurora and the uneasiness with which people viewed the aurora, ending with a tongue-in-cheek explanation of its origin.

The ballad is printed on the following pages in its entirety, courtesy of Dodd, Mead, and Co., New York.

THE BALLAD OF THE NORTHERN LIGHTS

One of the Down and Out—that's me. Stare at me well, ay,
  stare!
Stare and shrink—say! you wouldn't think that I was a mil-
  lionaire.
Look at my face, it's crimped and gouged—one of them death-
  mask things;
Don't seem the sort of man, do I, as might be the pal of kings?
Slouching along in smelly rags, a bleary-eyed, no-good bum;
A knight of the hollow needle, pard, spewed from the sodden
  slum.
Look me all over from head to foot; how much would you think
  I was worth?
A dollar? a dime? a nickel? Why, *I'm the wealthiest man on
  earth.*

No, don't you think that I'm off my base. You'll sing a different
  tune
If only you'll let me spin my yarn. Come over to this saloon;
Wet my throat—it's as dry as chalk, and seeing as how it's you,
I'll tell the tale of a Northern trail, and so help me God, it's true.
I'll tell of the howling wilderness and the haggard Arctic
  heights,
Of a reckless vow that I made, and how *I staked the Northern
  Lights.*

Remember the year of the Big Stampede and the trail of Ninety-
  eight,
When the eyes of the world were turned to the North, and the
  hearts of men elate;
Hearts of the old dare-devil breed thrilled at the wondrous
  strike,
And to every man who could hold a pan came the message, "Up
  and hike."
Well, I was there with the best of them, and I knew I would not
  fail.
You wouldn't believe it to see me now; but wait till you've heard
  my tale.

You've read of the trail of Ninety-eight, but its woe no man
  may tell;
It was all of a piece and a whole yard wide, and the name of the
  brand was "Hell."
We heard the call and we staked our all; we were plungers playing blind,
And no man cared how his neighbor fared, and no man looked
  behind;
For a ruthless greed was born of need, and the weakling went to
  the wall,
And a curse might avail where a prayer would fail, and the gold
  lust crazed us all.

Bold were we, and they called us three the "Unholy Trinity";
There was Ole Olson, the Sailor Swede, and the Dago Kid and
  me.
We were the discards of the pack, the foreloopers of Unrest,
Reckless spirits of fierce revolt in the ferment of the West.
We were bound to win and we revelled in the hardships of the
  way.
We staked our ground and our hopes were crowned, and we
  hoisted out the pay.
We were rich in a day beyond our dreams, it was gold from the
  grass-roots down;
But we weren't used to such sudden wealth, and there was
  the siren town.
We were crude and careless frontiersmen, with much in us of
  the beast;
We could bear the famine worthily, but we lost our heads at
  the feast.
The town looked mighty bright to us, with a bunch of dust to
  spend,
And nothing was half too good them days, and everyone was
  our friend.
Wining meant more than mining then, and life was a dizzy
  whirl,
Gambling and dropping chunks of gold down the neck of a
  dance-hall girl;
Till we went clean mad, it seems to me, and we squandered our
  last poke,
And we sold our claim, and we found ourselves one bitter morning—broke.

The Dago Kid he dreamed a dream of his mother's aunt who
  died—
In the dawn-light dim she came to him, and she stood by his
  bedside,
And she said: "Go forth to the highest North till a lonely trail
  ye find;
Follow it far and trust your star, and fortune will be kind."
But I jeered at him, and then there came the Sailor Swede to
  me,
And he said: "I dreamed of my sister's son, who croaked at the
  age of three.
From the herded dead he sneaked and said: 'Seek you an Arctic
  trail,
'Tis pale and grim by the Polar rim, but seek and ye shall not
  fail.' "
And lo! that night I too did dream of my mother's sister's son,
And he said to me: "By the Arctic Sea there's a treasure to be
  won.
Follow and follow a lone moose trail, till you come to a valley
  grim,
On the slope of the lonely watershed that borders the Polar
  brim."
Then I woke my pals, and soft we swore by the mystic Silver
  Flail,
'Twas the hand of Fate, and to-morrow straight we would seek
  the lone moose trail.

We watched the groaning ice wrench free, crash on with a hol-
  low din;
Men of the wilderness were we, freed from the taint of sin.
The mighty river snatched us up and it bore us swift along;
The days were bright, and the morning light was sweet with
  jewelled song.
We poled and lined up nameless streams, portaged o'er hill and
  plain;
We burnt our boat to save the nails, and built our boat again;
We guessed and groped, North, ever North, with many a twist
  and turn;
We saw ablaze in the deathless days the splendid sunsets burn.
O'er soundless lakes where the grayling makes a rush at the
  clumsy fly;
By bluffs so steep that the hard-hit sheep falls sheer from out the
  sky;
By lilied pools where the bull moose cools and wallows in huge
  content;
By rocky lairs where the pig-eyed bears peered at our tiny tent,
Through the black canyon's angry foam we hurled to dreamy
  bars,
And round in a ring the dog-nosed peaks bayed to the mocking
  stars.
Spring and summer and autumn went; the sky had a tallow
  gleam,
Yet North and ever North we pressed to the land of our Golden
  Dream.

So we came at last to a tundra vast and dark and grim and lone;
And there was the little lone moose trail, and we knew it for our
  own.
By muskeg hollow and nigger-head it wandered endlessly;
Sorry of heart and sore of foot, weary men were we.
The short-lived sun had a leaden glare and the darkness came
  too soon,
And stationed there with a solemn stare was the pinched, anaemic
  moon.
Silence and silvern solitude till it made you dumbly shrink,
And you thought to hear with an outward ear the things you
  ought to think.

Oh, it was wild and weird and wan, and ever in camp o' nights
We would watch and watch the silver dance of the mystic
  Northern Lights.
And soft they danced from the Polar sky and swept in primrose
  haze;
And swift they pranced with their silver feet, and pierced with a
  blinding blaze.
They danced a cotillion in the sky; they were rose and silver
  shod;
It was not good for the eyes of man—'twas a sight for the eyes
  of God

It made us mad and strange and sad, and the gold whereof we dreamed
Was all forgot, and our only thought was of the lights that gleamed.

Oh, the tundra sponge it was golden brown, and some was a bright blood-red;
And the reindeer moss gleamed here and there like the tombstones of the dead.
And in and out and around about the little trail ran clear,
And we hated it with a deadly hate and we feared with a deadly fear.
And the skies of night were alive with light, with a throbbing, thrilling flame;
Amber and rose and violet, opal and gold it came.
It swept the sky like a giant scythe, it quivered back to a wedge;
Argently bright, it cleft the night with a wavy golden edge.
Pennants of silver waved and streamed, lazy banners unfurled;
Sudden splendors of sabres gleamed, lightning javelins were hurled.
There in our awe we crouched and saw with our wild, uplifted eyes
Charge and retire the hosts of fire in the battlefield of the skies.

But all things come to an end at last, and the muskeg melted away,
And frowning down to bar our path a muddle of mountains lay.
And a gorge sheered up in granite walls, and the moose trail crept betwixt;
'Twas as if the earth had gaped too far and her stony jaws were fixt.

Then the winter fell with a sudden swoop, and the heavy clouds sagged low,
And earth and sky were blotted out in a whirl of driving snow.

We were climbing up a glacier in the neck of a mountain pass,
When the Dago Kid slipped down and fell into a deep crevasse,
When we got him out one leg hung limp, and his brow was wreathed with pain,
And he says: "Tis badly broken, boys, and I'll never walk again.
It's death for all if ye linger here, and that's no cursed lie;
Go on, go on while the trail is good, and leave me down to die."
He raved and swore, but we tended him with our uncouth, clumsy care.
The camp-fire gleamed and he gazed and dreamed with a fixed and curious stare.
Then all at once he grabbed my gun and he put it to his head,
And he says: "I'll fix it for you, boys"—them are the words he said.

So we sewed him up in a canvas sack and we slung him to a tree;
And the stars like needles stabbed our eyes, and woeful men were we.
And on we went on our woeful way, wrapped in a daze of dream,
And the Northern Lights in the crystal nights came forth with a mystic gleam.
They danced and they danced the devil-dance over the naked snow;
And soft they rolled like a tide upshoaled with a ceaseless ebb and flow,
They rippled green with a wondrous sheen, they fluttered out like a fan;
They spread with a blaze of rose-pink rays never yet seen of man.
They writhed like a brood of angry snakes, hissing and sulphur pale;
Then swift they changed to a dragon vast, lashing a cloven tail.
It seemed to us, as we gazed aloft with an everlasting stare,
The sky was a pit of bale and dread, and a monster revelled there.

We climbed the rise of a hog-back range that was desolate and drear,
When the Saiior Swede had a crazy fit, and he got to talking queer.
He talked of his home in Oregon and the peach trees all in bloom,
And the fern head-high, and the topaz sky, and the forest's scented gloom.
He talked of the sins of his misspent life, and then he seemed to brood,
And I watched him there like a fox a hare, for I knew it was not good.
And sure enough in the dim dawn-light I missed him from the tent,
And a fresh trail broke through the crusted snow, and I knew not where it went.

But I followed it o'er the seamless waste, and I found him at shut of day,
Naked there as a new-born babe—so I left him where he lay.

Day after day was sinister, and I fought fierce-eyed despair,
And I clung to life, and I struggled on, I knew not why nor where.
I packed my grub in short relays, and I cowered down in my tent,
And the world around was purged of sound like a frozen continent.
Day after day was dark as death, but ever and ever at nights,
With a brilliancy that grew and grew, blazed up the Northern Lights.

They rolled around with a soundless sound like softly bruiséd silk;
They poured into the bowl of the sky with the gentle flow of milk.
In eager, pulsing violet their wheeling chariots came,
Or they poised above the Polar rim like a coronal of flame.
From depths of darkness fathomless their lancing rays were hurled,
Like the all-combining search-lights of the navies of the world.
There on the roof-pole of the world as one bewitched I gazed,
And howled and grovelled like a beast as the awful splendors blazed.
My eyes were seared, yet thralled I peered through the parka hood nigh blind;
But I staggered on to the lights that shone, and never I looked behind.

There is a mountain round and low that lies by the Polar rim,
And I climbed its height in a whirl of light, and I peered o'er its jagged brim;
And there in a crater deep and vast, ungained, unguessed of men,
The mystery of the Arctic world was flashed into my ken.
For there these poor dim eyes of mine beheld the sight of sights—
That hollow ring was the source and spring of the mystic Northern Lights.

Then I staked that place from crown to base, and I hit the homeward trail,
Ah, God! it was good, though my eyes were blurred, and I crawled like a sickly snail.
In that vast white world where the silent sky communes with the silent snow,
In hunger and cold and misery I wandered to and fro.
But the Lord took pity on my pain, and He led me to the sea,
And some-ice-bound whalers heard my moan, and they fed and sheltered me.
They fed the feeble scarecrow thing that stumbled out of the wild
With the ravaged face of a mask of death and the wandering wits of a child—
A craven, cowering bag of bones that once had been a man.
They tended me and they brought me back to the world, and here I am.

Some say that the Northern Lights are the glare of the Arctic ice and snow;
And some that it's electricity, and nobody seems to know.
But I'll tell you now—and if I lie, may my lips be stricken dumb—
It's a *mine*, a mine of the precious stuff that men call radium.
It's a million dollars a pound, they say, and there's tons and tons in sight.
You can see it gleam in a golden stream in the solitudes of night.
And it's mine, all mine—and say! if you have a hundred plunks to spare,
I'll let you have the chance of your life, I'll sell you a quarter share.

You turn it down? Well, I'll make it ten seeing as you are my friend.
Nothing doing? Say! don't be hard—have you got a dollar to lend?
Just a dollar to help me out, I know you'll treat me white;
I'll do as much for you some day . . . God bless you, sir; good-night.

Most people visualize the earth as a small planet in the
endless black void of empty space. In reality, we are
surrounded by a dynamically changing region of charged
electrical particles and magnetic and electric fields. This
surrounding envelope of fields and particles
is called the magnetosphere.

15
The Magnetosphere

Magnus magnes ipse est globus terristris.
(The earth itself is a great magnet.)

William Gilbert
De Magnete, 1600

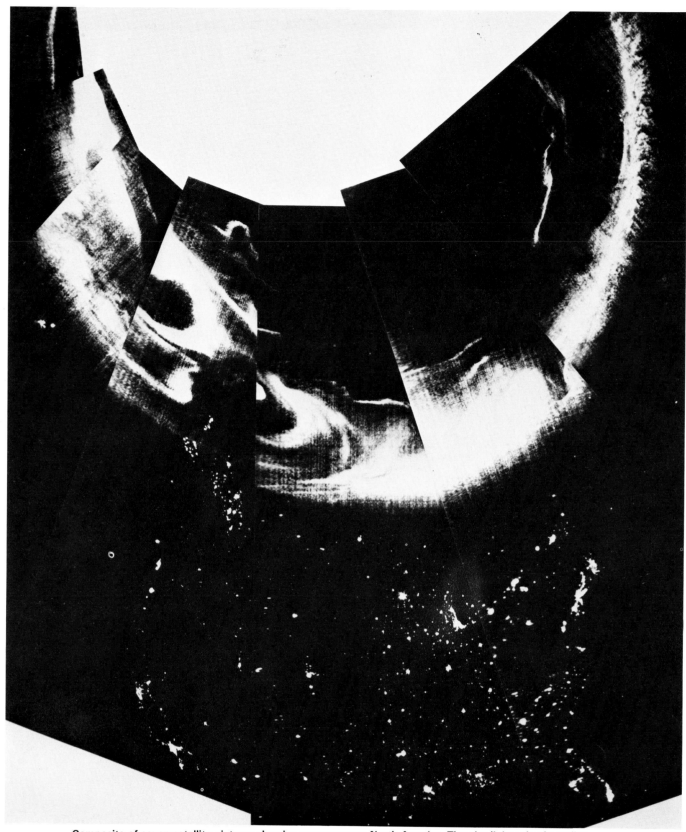

Composite of seven satellite pictures showing aurora across North America. The city lights clearly outline the continental United States.

The past 20 years have seen an explosion in the literature of research papers dealing with the aurora. A quadrennial report (147) on advances in auroral physics between 1971 and 1974 listed 425 research papers published in that time interval as well as three books and three documentary films. It is not possible or desirable to continue a historical path through such a dense jungle, and it would seem appropriate at this stage to simply present our present-day (1980) understanding of the auroral phenomenon.

Modern understanding of the aurora is tied up with our understanding of the magnetosphere as a whole. Even the word *magnetosphere* is relatively new in science (199) having been coined by Gold in 1958 to describe the region of near-earth space that is threaded by magnetic field lines linked to the earth and in which very hot or ionized gas dominates over the neutral atmosphere. It thus represents the outer limits of man's environment and is populated with ions and electrons originating in both the earth's atmosphere and the sun's atmosphere. It has become clear in recent years that the magnetosphere must be studied as a whole, as one single system of strongly interacting regions, none of which can be considered in isolation. Thus theoretical descriptions are almost prohibitively complicated, and a given observational manifestation such as the aurora is part of a complicated closed chain of cause and effect relationships. Roederer recently commented (395) that simplified models "are as unreal as would be the concept of a slightly pregnant woman."

That there should be an outflow of hot plasma from the sun (at least at particular times), a *solar wind*, had long been predicted theoretically. In 1931, Chapman and Ferraro (106) recognized that the earth's magnetic field would carve out a cavity in this flow. The solar wind may be viewed as pushing the earth's magnetic field, the compressed magnetic field in turn pushing an equal and opposite amount on the solar wind and halting its flow. Consequently, at a distance of some 40,000 miles, there is a transition region where the electrons and protons composing the solar wind can penetrate no closer to the earth and are deflected around the magnetic cavity. The simple dipole picture of the earth's magnetic field is considerably distorted by this compression on the sunward side and an elongation on the side away from the sun. This elongation results from particle "friction" with the magnetic field which allows the solar wind to drag the magnetic field lines into a cometlike tail that extends well out past the orbit of the moon (see pages 228–229).

The first evidence for the existence of a continuous solar wind (rather than an intermittent wind associated with unusual solar activity) came from comet studies. In 1950, Bierman's calculations (54) showed that pressure from the sun's light radiation was insufficient to explain the fact that comet tails always point away from the sun. He concluded that there must be a steady flow of particles from the sun to explain this observed orientation. (Birkeland had offered this same suggestion to explain comet tails as early as 1908, but in the intervening period, radiation pressure had become the accepted theory.)

The Chapman-Ferraro cavity was dramatically confirmed in 1961 when the *Explorer 14* satellite first crossed the boundary of the cavity on the sunward side of the magnetosphere. But the original theory could not explain the aurora because an essential ingredient, the interplanetary magnetic field, was not taken into account. The solar wind con-

William Gilbert 1540-1603
English physician and natural philosopher

Thomas Gold
Professor of Physics at Cornell University, New York

Definition of a *plasma:* A plasma is a fully ionized gas with equal numbers of positive and negative ions per unit volume. In terms of the solar wind and the magnetosphere, this means that the total number of positive ions (mostly protons) in any volume equals the total number of electrons.

L. Bierman
German physicist

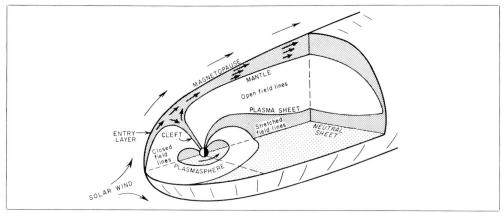

Plasma regions within the magnetosphere.

actions with the corotating neutral atmosphere, whereas the plasma sheet and mantle regions remain essentially fixed in space as the earth rotates within.

It is interesting to probe the reasons why magnetospheric plasma ends up in these curiously shaped regions; to illustrate some of the physical processes involved, we consider the possible fate of a solar wind electron or proton when it encounters the earth's magnetosphere. A large fraction of entering solar wind particles (though constituting only a few percent of impinging particles) gain access at the front of the dayside magnetosphere, though the dominant entry mechanism has not been satisfactorily identified. The magnetic cusp region, where magnetic field lines change from "closed" (joined to the other hemisphere) to "open" (swept into the magnetic tail) configuration, forms a natural access funnel for some of the recently entered plasma to flow down to the dayside atmosphere. These particles collide with the oxygen and nitrogen of the upper atmosphere and generate the dayside aurora, which extends at least 60° in longitude either side of the noon meridian. These dayside aurora never reach the spectacular brightness of the nighttime auroras. They are typically high aurora and hence red in color, as the precipitating particles only have enough energy to penetrate down to heights of about 150 miles, where the red spectral emission from atomic oxygen dominates. Because daytime aurora are directly linked with solar wind entry to our atmosphere, scientists maintain observatories at such inhospitable locations as the South Pole to monitor this weak red aurora (see page 147).

But much of the entry plasma streams along the open field lines along

(Left) A sketch of part of the vast current systems thought to exist in the magnetosphere. The circuit connecting the auroral zone is thought to be enhanced greatly during substorms.

(Right) A popularized sketch showing the 'funnel' whereby the solar wind particles may have direct access to the atmosphere on the dayside of the earth.

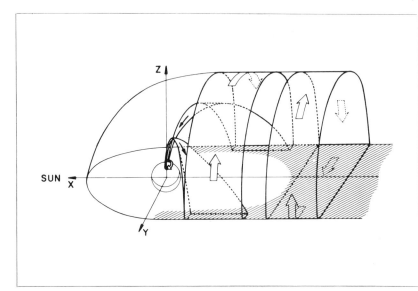

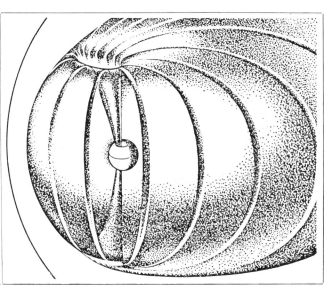

the flanks of the magnetic tail and forms the so-called mantle. During this journey the plasma is subject to the electric field generated by the dynamo action, which eventually forces it toward the central region of the tail, and the resulting particle accumulation forms the plasma sheet. There is probably also some solar wind access to the plasma sheet directly through the flanks of the downstream magnetosphere. A third possible source of plasma sheet particles may be the loss of the earth's own ionospheric particles along the open magnetic field lines from the polar cap regions.

An equilibrium situation is set up where the rate of build-up of the plasma sheet is balanced by loss from the plasma sheet. This loss proceeds via two paths: the electric field transports some particles back to the boundary of the magnetosphere (where they are lost back to interplanetary space), and some particles flow along the magnetic field lines which connect to the nightside auroral zone. This equilibrium flow to the auroral zones generates a continual, usually subvisual aurora called the "diffuse aurora"; this is not the bright and spectacular polar aurora of poets and legends, but, as we will see (page 224), it is probably the plasma sheet that is also the source of the brighter nighttime auroras. From the auroral point of view the plasma sheet is the most interesting region of the magnetosphere. It has rather well-defined boundaries: the inner (equatorward) boundary represents the transition between stretched (but still closed) field lines and open (nonconnecting) field lines, and the outer (poleward) edge locates at that region where the magnetospheric electric field pattern drives plasma back into inter-planetary space at a faster rate than can be replenished by solar wind entry.

Thus far we have described a model where magnetic field lines from the sun are imbedded in the solar wind and convected to the earth, where they merge with the earth's own magnetic field. This allows the solar wind to blow through the merging region and generate a giant voltage source that drives magnetospheric processes. Much of the current set up in the magnetospheric plasma mantle closes across the midplane of the plasma sheet in the magnetic tail, from the dawnside to the duskside, but a large-scale convective motion toward the earth is also set up. A portion of this gigantic current system closes across the auroral oval ionosphere, the circuit consisting of inward currents flowing down magnetic field lines to the morning polar ionosphere, connecting around the oval to the afternoon-evening sector, and then flowing back up the magnetic field lines (see page 220). It is this portion of the current system, closing through the resistive ionosphere, that is thought to delineate the auroral ovals in each hemisphere. And it is this tenuous connection, via magnetic field lines, that guides protons and electrons all the way from the sun's atmosphere to our own, where they collide with nitrogen and oxygen in the high atmosphere and so announce their arrival by the lights of the polar aurora.

But our story of the aurora is not yet complete. We have described the steady state or equilibrium situation which results only in dull and aesthetically uninteresting auroras. Under certain conditions, however, the interplanetary magnetic field appears to act as a trigger of a global instability called the *magnetospheric substorm*. The bright and spectacular auroras are the upper-atmosphere manifestations of such substorms.

The fourth Orbiting Geophysical Observatory ready for launch into polar orbit from the Western Test Range in California (July 1967). Many of the 20 experiments on board were specifically designed to study the aurora and its relation to precipitating particles and magnetic activity. The satellite operated successfully and greatly contributed to our understanding of the aurora.

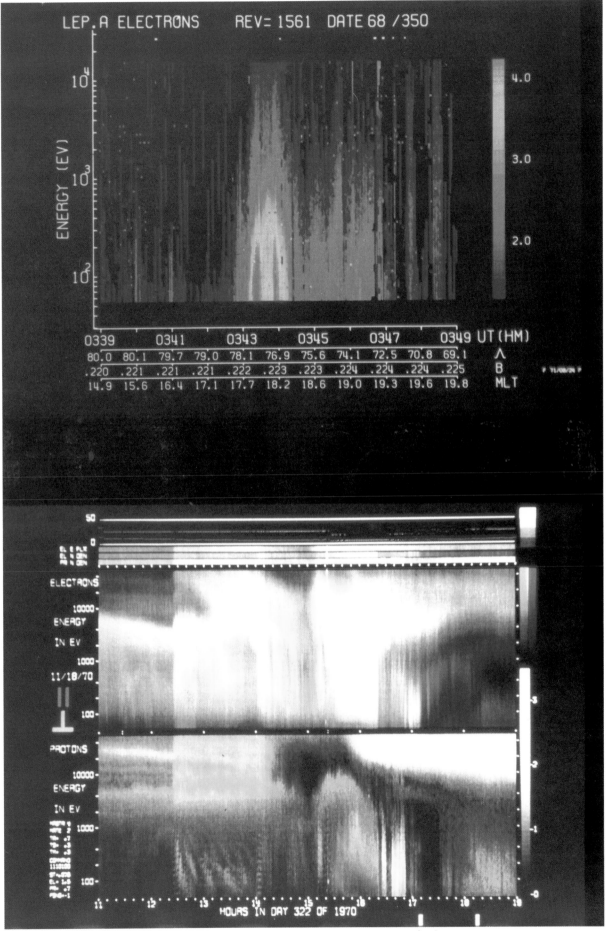

Examples of computer color-coding of satellite data on auroral particle fluxes. Hundreds of thousands of individual data points are presented on these pictures, which are reminiscent of the shapes and colors of the aurora itself.

(Top) A low-altitude polar satellite passes above an aurora and records the energy and flux of precipitating electrons.

(Bottom) 8 hours of measurements of protons and electrons 22,000 miles from the earth are shown here. The sudden vertical whitening represents substorm injection of fresh particles from the plasma sheet in the earth's geomagnetic tail; some of these particles precipitate down magnetic field lines to cause the bright auroras associated with these substorms.

The earth's magnetic field direction is from the north to the south, whereas that of the interplanetary field can take any direction. When the interplanetary field turns southward, many scientists believe that it can then more easily connect with the earth's field, which facilitates solar wind entry into the boundary layer and increases the rate at which polar cap field lines are dragged into the magnetic tail. This compresses magnetic field and plasma in the tail until a natural limit is reached beyond which the plasma sheet seems unable to tolerate further increases. This forced magnetic field line merging, or reconnection, results in a change in magnetic field configuration by forming a so-called neutral point in the magnetic field configuration near the center of the tail. Such a configuration is highly unstable and quickly relaxes to a more stable state. In this explosive process, which usually takes only 10–30 minutes, accumulated magnetic energy is somehow converted into kinetic energy of the plasma sheet particles which are "squirted" away from the neutral region. Particles on the earthward side of this unstable region may suddenly find themselves rushing earthward at 1,000 miles per second, whence they can either precipitate into the atmosphere or be injected closer toward the earth to join the Van Allen radiation belt.

It is these energized particles, ejected from the plasma sheet because of a magnetically unstable region in the magnetic tail, that generate the fascinating display of bright luminosity that has fascinated man since he first gazed into the polar sky.

The above scenario has been the most popular among scientists in recent years, but there is certainly no universal agreement. Many doubt that magnetic field merging is as important as is implied in this explanation; others believe much of the energization of auroral particles results from large electric fields much closer to the earth than the plasma sheet. And the triggering process for the substorms cannot be as simple as the interplanetary field turning southward to align with the earth's field, as substorms are also observed to occur with the interplanetary field pointing northward.

In fact the most recent research on these topics seems to indicate that it may be possible to understand the gross worldwide features of substorms as being directly driven by external changes in the solar wind parameters (the magnitude of the solar wind magnetic field and its direction and the speed of its flow). Conceivably, "macrostructure" changes

Simplified sketch showing the chain of events that some scientists believe describe substorm process.

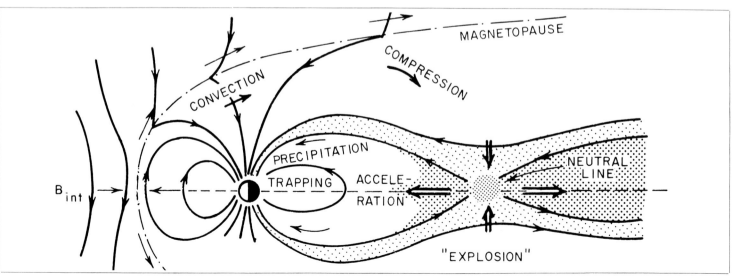

224

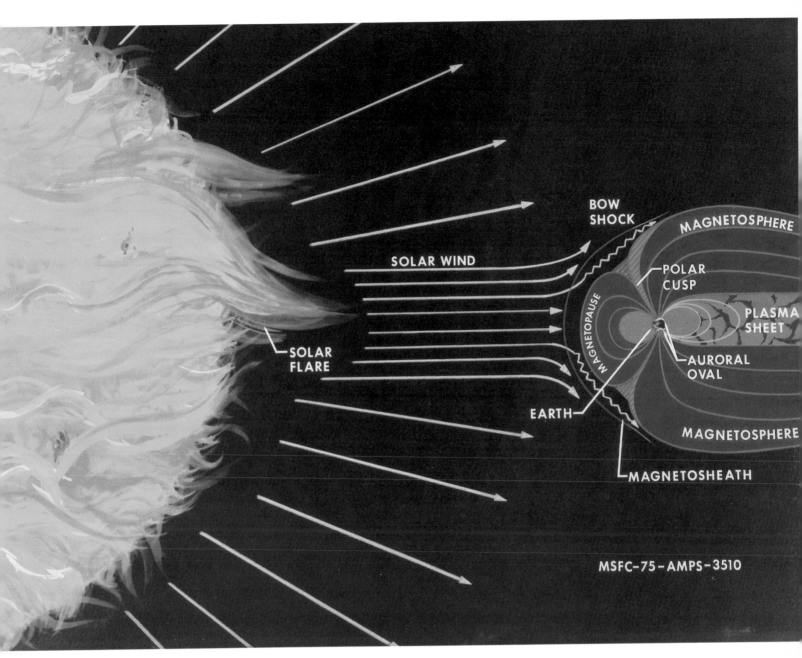

Sketch (not to scale) emphasizing the controlling influence of the sun on the region around our planet.

such as these can affect the large-scale magnetospheric current systems, which in turn affect the location of the auroral oval. But the dynamic "microstructure" of the aurora in various localized regions of the oval must then develop more or less unpredictably from local perturbations in parameters much closer to the earth, for example, in ionospheric conductivity and low-altitude electric fields. The situation is analogous to that of weather prediction, where forecasters achieve mixed success in trying to predict the details of weather behavior at a given place from a knowledge of imminent passage of a large-scale weather system.

Understanding the substorm process has been the principal goal of space and auroral research for the past decade. But a satellite making particle measurements in space is just a pinpoint "like an ant walking around the Rocky Mountains trying to understand their geology" (175). And as satellites move so rapidly through space, one is never sure whether measured changes are localized and spatial or widespread and temporal. The practical difficulty is to obtain representative observational coverage, and ground-based auroral measurements suffer from similar limitations. Substorm auroras cover vast areas of inhospitable polar regions. Early research attempted to piece together the worldwide picture from scattered all-sky cameras; a major advance came in the early 1970's with the availability of pictures of the auroral oval taken from the *Isis* spacecraft (300) and from an Air Force satellite (396). One photograph of a 2,000-mile-wide strip was obtained from each orbit, and successive orbital passes (separated by approximately 90 minutes) covered adjoining 2,000-mile strips (see page 216). Thus the auroral development over fairly large areas at particular stages of substorm development could be studied. But one of the most exciting goals of auroral research is to be able to photograph the complete auroral oval over extended time periods with good time resolution, so that the complete spatial and temporal development of substorms can be observed. An imaging experiment is presently planned for an eccentric orbit spacecraft, which hopefully will achieve this goal early in the 1980's.

It is sobering to realize that the auroral light represents only about 4% of the energy that precipitates into the atmosphere, the remainder heating the neutral atmosphere and causing ionization, ultraviolet and infrared emissions, and various radio and X ray emissions. And of the total energy brought to our magnetosphere by the solar wind, only a few percent end up as kinetic energy of precipitating particles. Thus the thousands of miles of aurora that stretch around the polar regions in both hemispheres only represent perhaps one tenth of 1% of the vast energy source that drives magnetospheric processes.

Even so, the rate of auroral energy release in the upper atmosphere during a moderate substorm (of the order of 100 million kilowatts) is comparable to the total power-generating capacity of all the man-made power plants in all the countries of the earth. As this energy is dispersed over such vast areas of the earth and at inaccessible heights, there is no possibility that this natural power source could ever be harvested by man.

The whole complex system can be likened to a television set, where the solar wind can be thought of as analogous to the flow of electricity from the wall socket into the television set, where the electrons that compose the electric current are emitted into the television tube by an "electron gun." (This electron gun effectively "boils" the electrons off a

Owen Garriott
Scientist and astronaut

On September 3, 1973, the sun erupted in violent activity. Two days later, solar wind particles arrived at the earth, and as Skylab passed south of Australia, astronaut Owen Garriott took this picture of an aurora that resulted. Also seen is a thin layer of airglow viewed 'edge on' and seen as a narrow band parallel to the earth's limb.

hot filament and then accelerates them toward the television screen.) Electric and magnetic fields deflect and guide the electrons to generate the shapes that compose the television picture. Our atmosphere is analogous to the television screen. Both glow where the electrons impinge after they have been subjected to accelerations and deflections by magnetic and electric fields between the source and the screen. The polar atmosphere is like a giant television screen where we see the aurora, the only visible manifestation of processes taking place in the magnetosphere. It is these processes, which have their analogy in the complex electronics of the television set, that we are trying to understand. Just as watching a television program tells us little about how the television set works, studying the aurora will not reveal the secret of the magnetospheric electronic circuits. Comprehensive measurements throughout this vast and complex system must be obtained and pieced together to derive a satisfactory understanding.

The space age began about 20 years ago with the launch of *Sputnik I* and *Explorer I*, and during that time we have become increasingly aware of how important this near-space region is to man and his interaction with his environment. The initial exploratory stage of magnetospheric physics is over, but we still do not understand many of the details of pro-

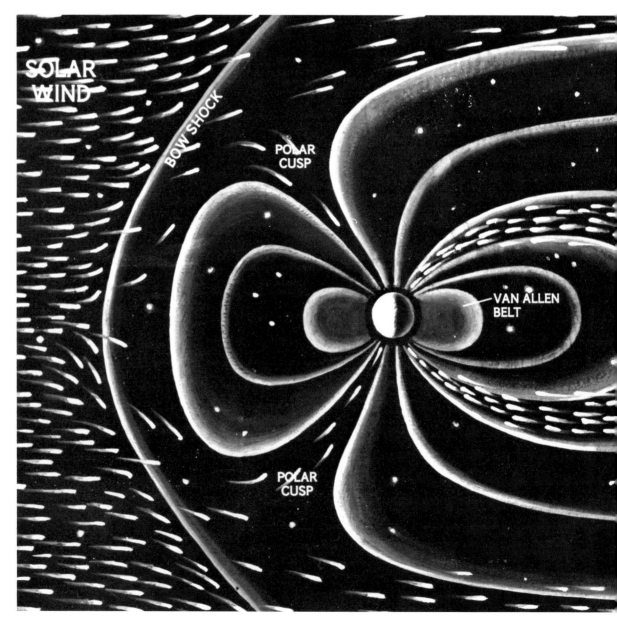

cesses whereby the magnetosphere affects the lower atmosphere where we live. Future experiments must be planned more carefully and rigorously and will involve coordinated satellite and ground-based observations as well as direct perturbation of the system.

The technical needs of our increasingly complex society have required utilization of larger portions of the space around the planet, including the placement of sophisticated hardware into that space. Such uses have been largely confined to travel and communications, but the future may see space-based solar power become a reality. As technologies change, we must understand how these changes might affect our environment, and to do that we must fully understand all the interactive components of the system. It is considerations such as these that ensure that magnetospheric and auroral research will be active fields of science for many years to come.

This sketch by Pierre Mion (for *National Geographic*) shows probable entry paths of solar wind particles to the magnetosphere. Some particles penetrate the bow shock, and the remainder are diverted around the magnetospheric cavity. Of those particles penetrating on the dayside, some find direct access to the dayside atmosphere through the polar cusp and generate the dayside aurora. Others flow around the earth along the magnetopause and a fraction of these slowly flow inward to add to the plasma sheet far behind the earth. It is the return flow of plasmasheet particles toward the earth that is important in generating the nighttime auroras. As they approach the earth, dynamic electric fields and instability regions are thought to energize them further, resulting in the varying colors and patterns of the aurora borealis and aurora australis.

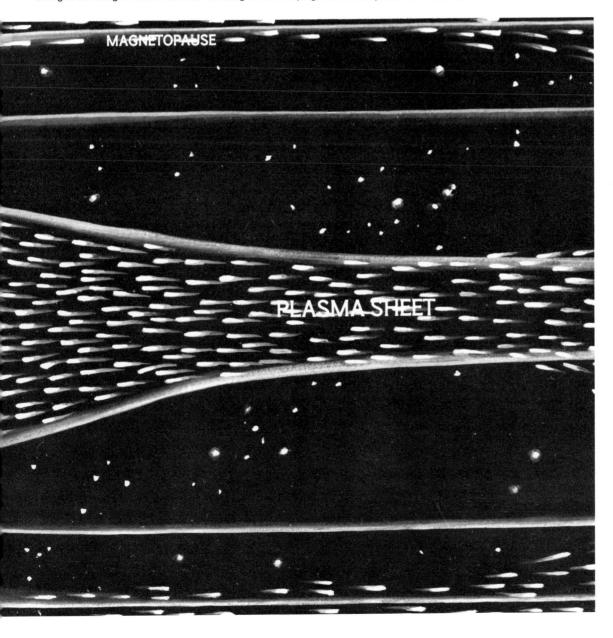

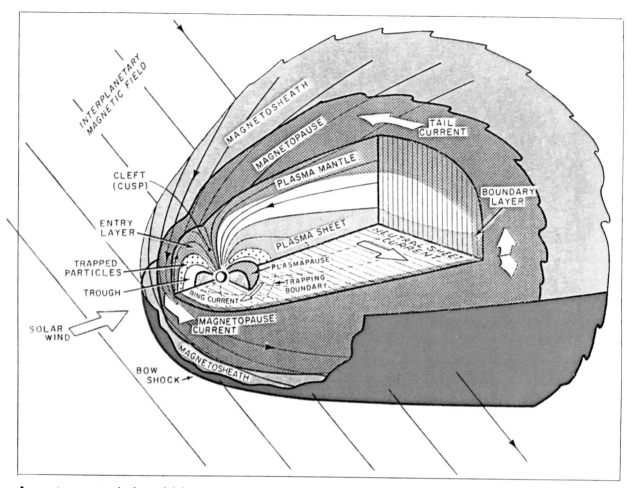

A recent magnetospheric model drawn by Heikkila, showing just how complex and sophisticated our current picture of the magnetosphere has become.

16
Man-Made Aurora

Do I dare
Disturb the Universe?

T. S. Eliot
The Lovesong of J. Alfred Prufrock, 1917

Man masters nature not by force but by understanding.
That is why science has succeeded where magic failed:
because it has looked for no spell to cast on nature.

Jacob Bronowski
Science and Human Values, 1956

This photograph was taken about 3 minutes after the release of a shaped-charged barium cloud from Tonopah, Nevada, in May 1978 (Los Alamos Scientific Laboratory 'AveFria' Program). The cloud was at a nominal height of 200 kilometers and quickly developed this beautiful auroralike striated structure.

One topic that has not been covered in the preceding chapters is that of man's attempt to generate model, or artificial, auroras. This was a deliberate omission, as the story covers the time frame of many of the earlier chapters and needs a knowledge of the last chapter on the magnetosphere to put it into proper perspective.

The mid-eighteenth century saw many new theories suggested to explain the aurora, but the problem was how to resolve the possibilities. Essentially all that was available to help the scientist was visual observations. It was not possible to bring the aurora to the laboratory for study nor to take instruments into the aurora. But then the discovery of electricity and electric discharges offered a new avenue of investigation; auroralike effects could be produced in an evacuated glass tube in the laboratory. Could the scientist in effect make his own aurora in a bottle? Many thought so, and scientists began to examine the properties of the "electric fluid" in discharge tubes and to try to relate them to natural aurora.

It appears that a Mr. Hawksbee in England was the first to show how electricity passing through a highly rarefied atmosphere (of an evacuated glass tube) gave a light whose appearance resembled the aurora (90). In 1750, Canton reported (89) on similar experiments which he related to the mechanism causing thunder and lightning; but he concluded with the query,

> Is not the aurora borealis, the flashing of electrical fire from positive, towards negative clouds at a great distance, through the upper atmosphere, where the resistance is least?

The belief that the light given out in various laboratory discharge tubes was analogous to the auroral light must have been almost universal in the late 1800's, for J. Rand Capron (90) in his textbook *Aurora: Their Characters and Spectra*, published in 1879, commented that

> With many of us (at least it was so in my case) our first viewed Aurora have been artificial ones, devised by electricians and having their locus at the Royal Polytechnic in Regent Street or in some science lecture-room. The effects in these cases are produced in tubes nearly exhausted by means of an air pump, and then illuminated by some form of electric or galvanic current.

A more elaborate demonstration (278) to create artificial aurora was devised in the 1870's by Professor Lemström in Finland in support of his theory that the aurora was produced by electric currents passing through the atmosphere. An "electrical machine" was connected to a copper sphere (representing the earth) above which were suspended (at some distance) a number of discharge tubes, the top ends of which connected back to the electrical machine (see page 235). When the machine was set in operation, a current passed through the air to the discharge tubes which became luminous. It was explained that electrical currents could pass unobserved in the air at atmospheric pressure but illuminated a rarefied atmosphere; this supported the contention of Swedish scientists that electrical currents emanating from the earth and passing into the upper atmosphere produce auroras.

De la Rive contrived a similar demonstration (135) about the same time but with the addition of cylindrical magnets projecting from the model earth at the poles (see page 235). The resultant discharges were claimed to "faithfully reproduce the aurora borealis and aurora australis with

John Canton 1718-1772
English scientist

J. Rand Capron 1829-1888
Solicitor and part-time scientist
Fellow Royal Astronomical Society

Selim Lemström 1838-1904
Professor of Physics at
University of Helsinki, Finland

their attendant phenomena." Both Lemström's and de la Rive's apparatus were exhibited to the general public at the Science Museum in London as part of the Scientific Loan Collection of 1876.

Less sophisticated "artificial auroras" were reported by M. Planté in 1878: a negative lead from a powerful battery was inserted in a vessel of salt water, and when the positive lead was then inserted near the wall of the same vessel, effects were produced which were considered to strongly resemble the aurora (370) (see opposite).

Capron in England was also studying the properties of various electric discharges, especially the effect of magnets on their shape and form. Many auroralike structures could be produced, and Capron examined the spectra from such tubes for comparison with the auroral spectra. He concluded (90) that

Apart from the spectroscopic questions involved, the oldest and most received theory of the aurora—that of its being some form of electric discharge in the more rarefied regions of the atmosphere—seems to hold its own . . . As the general result of spectrum work on the aurora up to the present time, we seemed to have quite failed in finding any spectrum which, as to position, intensity and general character of the lines, well coincides with that of the aurora . . . The whole subject may still be characterized a scientific mystery.

Capron's experiments conducted to test, in connection with the aurora, the action of a magnet on an electric glow in vacuo and on a spark at ordinary pressure. The results proved 'assuming the aurora to be an electric discharge, the great influence the magnetic forces may exercise on the colours, forms, motions, and probably the spectrum also of that phenomenon.'

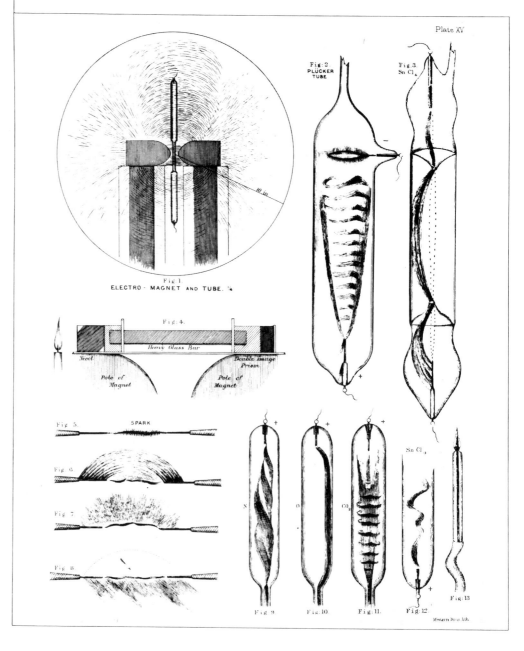

234

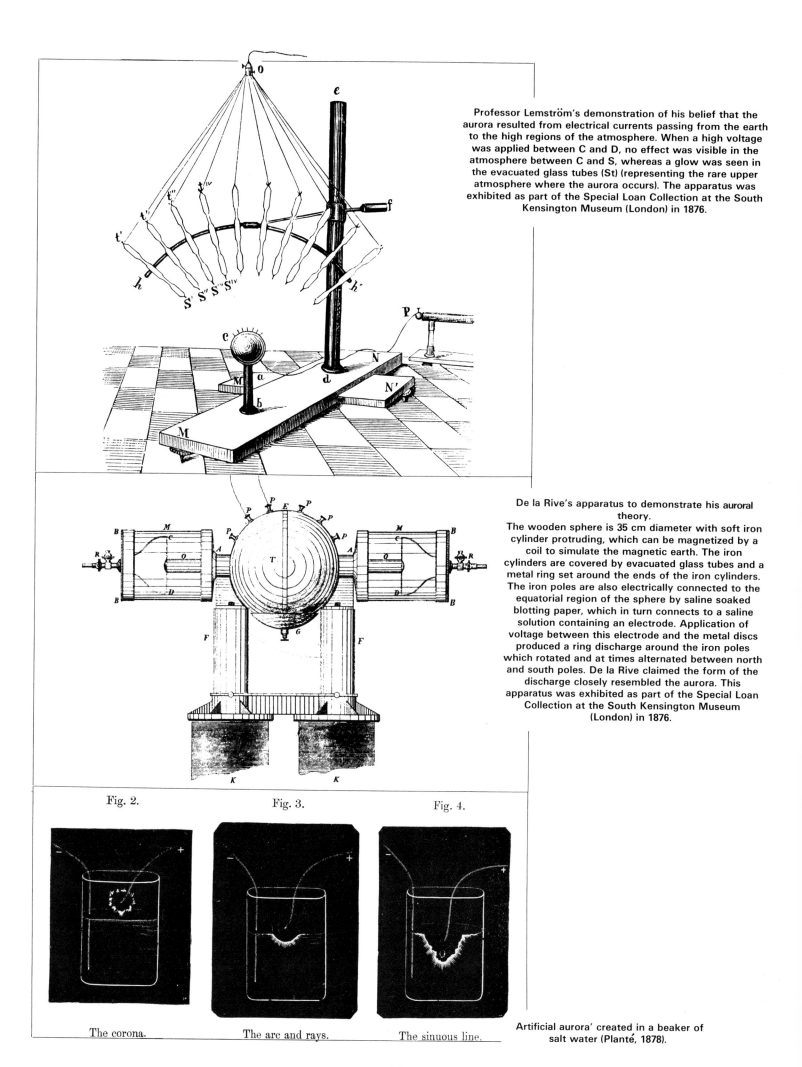

Professor Lemström's demonstration of his belief that the aurora resulted from electrical currents passing from the earth to the high regions of the atmosphere. When a high voltage was applied between C and D, no effect was visible in the atmosphere between C and S, whereas a glow was seen in the evacuated glass tubes (St) (representing the rare upper atmosphere where the aurora occurs). The apparatus was exhibited as part of the Special Loan Collection at the South Kensington Museum (London) in 1876.

De la Rive's apparatus to demonstrate his auroral theory.
The wooden sphere is 35 cm diameter with soft iron cylinder protruding, which can be magnetized by a coil to simulate the magnetic earth. The iron cylinders are covered by evacuated glass tubes and a metal ring set around the ends of the iron cylinders. The iron poles are also electrically connected to the equatorial region of the sphere by saline soaked blotting paper, which in turn connects to a saline solution containing an electrode. Application of voltage between this electrode and the metal discs produced a ring discharge around the iron poles which rotated and at times alternated between north and south poles. De la Rive claimed the form of the discharge closely resembled the aurora. This apparatus was exhibited as part of the Special Loan Collection at the South Kensington Museum (London) in 1876.

Fig. 2.   Fig. 3.   Fig. 4.

The corona.   The arc and rays.   The sinuous line.

Artificial aurora' created in a beaker of salt water (Planté, 1878).

235

Meanwhile, Professor Lemström in Finland was still busy with auroral experiments, and in 1882–1883 he claimed to have created a full-scale artificial aurora (279). On top of the Mount Oratunturi, near Sodankyla, he erected his *ustromnings*, or discharge apparatus, consisting of copper wire spiral covering 3,000 square feet with tiny points of iron soldered every 18 inches, connected to a platinum disc buried at the base of the mountain. At night, Lemström said the apparatus was "surrounded by a faint, yellow-white luminosity, which, when examined with the spectroscope, gave—though faintly—the same spectrum as the Aurora Borealis." He then concluded that the aurora occurred very near the surface of the earth. Tromholt was cautious of Lemström's claims and decided to test them himself by erecting a similar apparatus on the top of Mount Esja in Iceland (472). He found no trace of artificial aurora "even though Iceland was nearer the auroral zone."

These experiments excited considerable interest in the scientific world, and in 1884, Vaussenat, the founder of the observatory on the Pic du Midi in France, erected an apparatus at 12,440 feet, similar to Lemström's but of even larger dimensions. The apparatus was supported by 200 posts, covered 7,000 square feet, and contained 14,000 points soldered to the wire. During the 10 months of operation the luminous rays claimed by Lemström were never seen (though it is not clear why such auroral effects were expected in France). The sole result (481) was to

. . . produce in the neighborhood of the wire violent electric discharges, which caused serious danger to the observers . . . Once, when near the apparatus, Vaussenat had his eyebrows burnt, the skin of his face was scorched, his clothes singed, and the spring of his watch affected.

Apparatus erected by Lemström on the top of a mountain in northern Finland. He claimed that the device produced artificial aurora.

Vaussenat's discharge apparatus, erected on the Pic du Midi in 1884, to test Lemström's claim of having created artificial aurora.

Celetin X. Vaussenat 1831-1891
French scientist

Various connections that were being established at this time between events on the sun and auroral events on the earth led to suggestions that the sun was the source of auroral particles. Becquerel [50] believed that sunspots were cavities by which hydrogen and other substances escaped from the sun and that the hydrogen would take with it positive electricity which would spread in interplanetary space to the earth itself. Goldstein [200] in 1881 adopted a similar view, but considered the solar terrestrial currents to consist of cathode rays.

Birkeland took up this argument [56] in the 1890's and began his well-publicized terrella experiments. In his laboratory in Oslo he built a model of the complete sun-earth system, as has already been described (see page 132). An electron gun fired electrons toward a magnetic model of the earth, impinging on the sphere in various regions, which were made visible by fluorescent paint. In this way he could simulate many features of the known auroral distributions.

Since Birkeland's pioneering work, there have been many terrella experiments set up in laboratories around the world to try to study the problem of the interaction between solar particles and the magnetosphere. Some of these model experiments have aimed only at the study of certain details of the problem, while others have attempted to simulate the entire phenomenon.

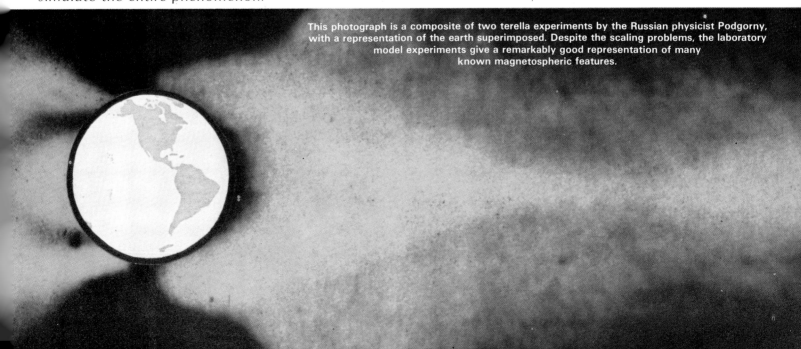

This photograph is a composite of two terella experiments by the Russian physicist Podgorny, with a representation of the earth superimposed. Despite the scaling problems, the laboratory model experiments give a remarkably good representation of many known magnetospheric features.

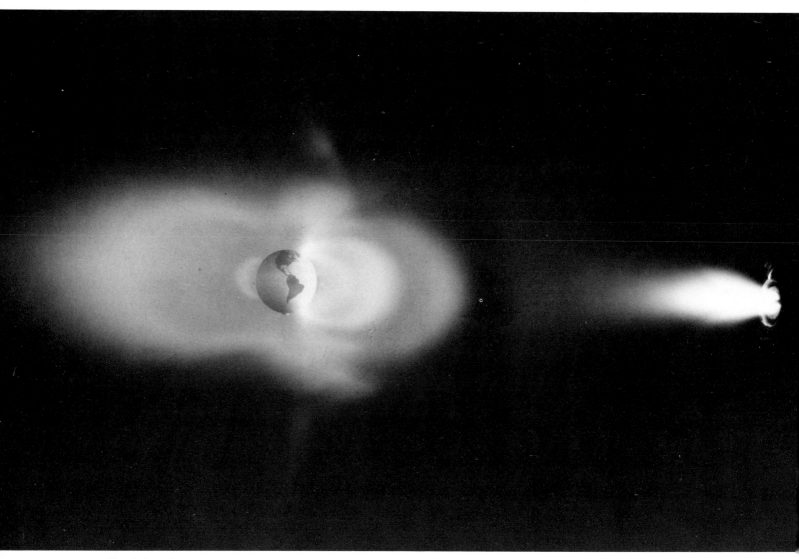

Simulation of the solar wind interacting with the earth's magnetic field. The solar wind is simulated by a plasma thruster (right). Plasma speed, magnetic field strength, and particle densities do not reproduce actual conditions in space but are adjusted so that radiation belts that are created are roughly to scale.

With increasing experimental and theoretical sophistication, the problem of meaningful scaling has become recognized (57). For example, it is not technically feasible to duplicate the very low particle densities of outer space in the laboratory. If the speed of electrons in the model experiment is the same as that in interplanetary space, the magnetic field strength for the model earth must be much greater than the real earth in order to stop the electrons within the dimensions of a laboratory vacuum tank; yet, if the electron speeds are not the same as those in outer space, the whole experiment may be meaningless because interactions between particles depend on their speed.

It has been shown that model experiments that deal with the entire space of interest cannot be scaled properly in all respects and consequently such experiments are of questionable significance. Despite this, there have been some impressive model experiments in modern times that give an uncanny resemblance to the particle distributions that we now know (from satellite measurements) to characterize the mag-

Photograph of the August 1, 1958, high-altitude nuclear explosion 'Teak' near Johnston Island in the Pacific. This photograph was taken from Hawaii, 1300 km from the explosion. The auroral arc extending southwards from the explosion is generated by fast electrons guided along the magnetic field line. A crimson-colored aurora was also seen from Apia in the southern Pacific, which is the southern end of this same magnetic field line.

netosphere. Perhaps the best example of this is the model experiment by the Russian physicist Podgorny (373); it is shown in the illustration on page 237. Certain aspects of the magnetospheric configuration clearly show through despite the unrealistic scaling involved; but in general, we cannot claim that these experiments will elucidate problems of the earth's magnetosphere. However, they certainly advance understanding of the confinement of plasmas in magnetic fields and so may well contribute to understanding laboratory plasma machines that attempt controlled thermonuclear fusion—a possible future source of unlimited and inexpensive energy.

Man's most spectacular artificial aurora occurred over the Pacific island of Samoa (13.8°S) on August 1, 1958, at 1051 GMT and lasted some 14 minutes. Aurora australis had been reported from this near-equatorial location only once before (19), in association with the severe magnetic storm of May 13–16, 1921. The violet, red, and green rays

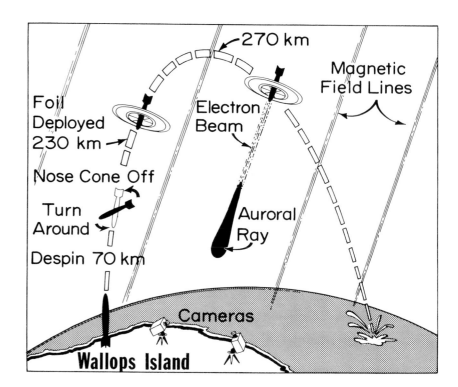

Schematic drawing of the first controlled 'man-made aurora' experiment. A rocket launched from Wallop's Island, Virginia, carried an electron accelerator that fired a beam back at the earth's atmosphere. The extremely weak auroral ray thus generated was detected by a high-sensitivity television camera; it is seen as the streak down the right-hand side of the picture superimposed on a dense field of (subvisual) stars.

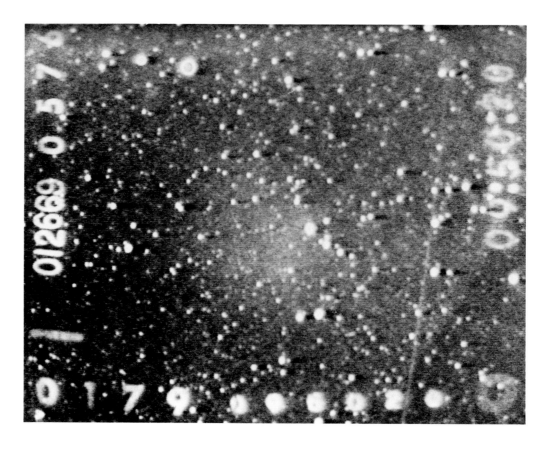

observed from Samoa in 1958 were at first thought to be another rare display of aurora australis, but there was no corresponding worldwide magnetic disturbance. It was later revealed that American scientists had exploded a nuclear bomb high in the atmosphere over Johnston Island (19.0°N). Johnston Island is at the other end of the magnetic field line that passes near Samoa. The "aurora" was man-made (126) and resulted from energetic particles generated in the explosion being guided by the magnetic field to the southern hemisphere location.

A more controlled attempt by man to make his own aurora took place on January 26, 1969, at Wallop's Island, Virginia. This was the date of the first rocket launching of an electron accelerator high above the atmosphere to shoot electrons back toward earth and thus simulate the auroral bombardment. The brief bursts from the electron gun gave weak flashes of auroral light at heights of approximately 65–80 miles which were recorded by sensitive television cameras (229). Similar experiments have been carried out in the 1970's to further our understanding of how such beams of electrons interact with the atmosphere to cause the auroral light and under what conditions (speed, density, and duration) the beams might become unstable and perhaps to explain some of the details of complex auroral structure.

The 1980's will see man-made auroral experiments continue from the Space Shuttle. The very first shuttle flights are expected to include a large electron gun to fire intense beams of electrons down into the atmosphere. Also on board will be sensitive television cameras to analyze the auroral light generated below the vehicle. But there are limits as to just how large and how bright an aurora man can make in this way. For every electron fired away from the shuttle, one must be collected from the immediate environment, otherwise the whole vehicle would quickly acquire a high positive charge that would pull the electrons back as soon as they left the gun.

And when we consider that the power deposited worldwide in a good aurora exceeds the entire power-generating capacity of all countries of the world combined, it is clear that puny man can never hope to assemble in space the hardware to compete with nature. Some scientists hold, however, that man might be able to perturb or trigger the natural magnetospheric environment to dump considerable energy of its own, and in that way we could turn on auroras on command. Some experiments to that end are already underway, but the general feeling among scientists is that it is of academic rather than of any practical interest.

The aurora has always been associated with the cold bleakness of the polar regions. On rare and memorable occasions she ventures forth to an admiring world. It would seem a pity if we were to ever change that.

Fred's auroral nursery.

Energy of the Aurora

There is enormous energy associated with strong substorm aurora, some 1,000 million kilowatts. This represents about 10 times the electrical power consumption of the United States. Unfortunately, this energy is spread over such a large area of the earth's surface and at such inaccessible heights that it would never be practical to harness auroral energy for man's use.

Such problems did not deter some nineteenth century speculation on the use of the aurora's energy. In his book *The Voyage of the Vivian to the North Pole and Beyond,* T. W. Knox has Dr. Tonner explain current ideas about the aurora to his two youthful charges, George and Fred. The boys then make the following speculation (530):

"We are getting into deep water," said George, "or rather we should be if there was less ice about us. When we have time to spare we will set about devising a machine whereby the electricity of the aurora borealis may be harnessed, and made to do duty in a practical way. We will make it run the dynamos to supply our houses and streets with electrical light; it shall propel our machinery, and thus take the place of steam; it shall be used for forcing our gardens, in the way that electricity is supposed to make plants grow; and it shall develop the brains of our statesmen and legislators, to make them wiser and better and of more practical use than they are at present. Hens shall lay more eggs, cows must give cream in place of milk, trees shall bear fruit of gold or silver, tear-drops shall be diamonds, and the rocks of the fields shall become alabaster or amber. Wonderful things will be done when we get the electricity of the aurora under our control."

"Yes," responded Fred, "babies shall be taken from the nursery and reared on electricity, which will be far more nutritious than their ordinary food. When the world is filled with giants nourished from the aurora, the ordinary mortal will tremble. We'll think it over, and see what we can do."

And with this cautious suggestion the conversation was changed to a more commonplace topic.

T. W. Knox
*Voyage of the Vivian to the North
Pole and Beyond*

242

17
What Now?

Short of compulsory humanistic indoctrination
of all scientists and engineers, with a 'Hippocratic
oath' of never using their brains to kill people,
I believe the best makeshift solution at present
is to give the alpha-minuses alternative outlets
for their dangerous brain-power, and this may
well be provided by space research.

Dennis Gabor
Inventing the Future, 1963

We all recognize that the era of exploration
in the near-earth space environment is over.
However, the era of understanding the behavior
of that environment has just begun.

Donald J. Williams
Guest Editorial
EOS, 1975

Thou sun, of this great world both eye and soul.
John Milton
Paradise Lost V 1, 171 (1667)

The sun is the ultimate source of energy and probably of the atomic particles that generate the aurora polaris.
This photograph was taken through a hydrogen α filter with the Space Environment Services telescope at
Boulder, Colorado, as the sun set behind the Flatirons, foothills of the Rocky Mountains.
The hydrogen α filter reveals sunspots, dark filaments and bright plages (December 1968).

244

Auroral physics has reached a state of coherence and understanding that makes it timely to ask the question, where do we go from here? Some feel that auroral and magnetospheric physics are in abatement, stemming not so much from the current decline in research funding as from a kind of intellectual irresolution, perhaps because the great age of space exploration and discovery has ended. This feeling is reminiscent of the low spirits of physicists around 1895:

> It was believed by all (or nearly all) that the possible great discoveries had been made. Professors of physics warned their students that other lines held out more promise, and that the future of physics was to be one of residuals, second order effects, and the never-ending quest for one more decimal place.
>
> Heyl, *Physics,* 1964

Twentieth century physics testifies to the fleeting nature of such ill-founded pessimism.

Similarly, there is much to be done in auroral physics. There are many crucial questions to be answered. We would like to know what factors facilitate solar wind entry into the magnetosphere. Is the merging rate, the joining of interplanetary and earth magnetic field lines, the most important controlling influence? Certainly, the direction of the interplanetary field is important, but it does not appear to be the whole story. And what exactly is the entry mechanism? Various diffusion, turbulence, and drift mechanisms have been suggested.

What is it that triggers substorms? If there was a simple upper limit to the storage capacity of the plasma sheet beyond which explosive relaxation occurred, then we would expect all substorms to be approximately the same strength. But substorms vary from the barely detectable to the spectacular, so what other factors are involved? What is the acceleration mechanism that operates on the particles "squirted" out of the plasma sheet during substorms? The particles that collide with the upper atmosphere have many times the energy that they had in the plasma sheet, and there is no accepted theory on how excess magnetic energy is transferred to kinetic energy of the particles. Perhaps the main energization occurs much closer to the earth by low-altitude electric fields. Indeed, the very source of some of the auroral particles may not be the plasma sheet at all but could be energization of ionospheric particles.

And what causes the fine structure, the beautiful swirls and loops of the aurora, and the rapid movements of these structures? No corresponding structure is evident in the particle flows measured by satellites in the inner part of the plasma sheet, so it seems that low-altitude mechanisms are involved. This speculation is supported by rocket experiments that have artificially injected ionized plasma (barium ions) into the high ionosphere; the resultant glowing cloud of plasma quickly develops a fine-striated structure similar to that seen regularly in auroras. There have been almost as many different types of plasma instabilities proposed to explain detailed auroral structures as there are interested theorists, and indeed, many different mechanisms probably operate at different times. Recent years have seen active probing of the ionosphere via intense radio beams directed from the ground, by rocket- and satellite-launched barium clouds, and by beams of electrons fired into the atmosphere by rocket-borne electron accelerators; all were conceived to answer questions concerning plasma instabilities that could generate auroral structures.

A barium cloud near 200 km, photographed soon after launch. The ionized barium ions (blue) have been moved away from the circular neutral cloud (green) by the electric field near the auroral zone. Ray structure typical of natural aurora has also developed.

LORO   1506:36 UT   OCT. 18, 1972

E+ 14 min

E+ 24 min

7000
6000
7200
5000
4000
MAG EQ
3000

61872

61872

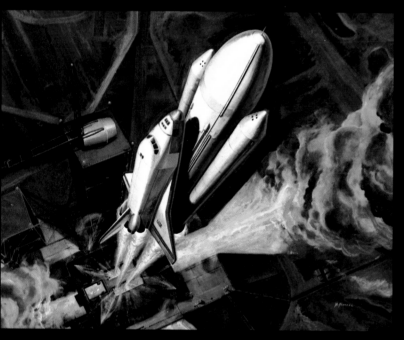

Auroral research used to be inexpensive; Kristian Birkeland privately funded his auroral expeditions. But in the 1960's and 1970's this has certainly not been the case—our more sophisticated experiments have become progressively more expensive. And John Citizen might rightfully question what is in it for him, as it is his tax dollars that support this research. In this respect it is appropriate to draw analogy from the following story. The physicist Robert Wilson was defending the need for a new particle accelerator before a U.S. Senate subcommittee. Senator Pastore asked if the machine would bring anything to help defend the country against its enemies. Wilson replied, "Absolutely not." Pastore then asked, "Well, what then is the use of it?" And Wilson said, "In the sense this new knowledge has nothing to do directly with defending our country except to make it worth defending."

It is interesting to observe that many of the most exciting experiments of the late 1970's were achieved from smaller and relatively inexpensive satellites, quickly assembled by mainly industrial research groups, and orbited in regions of near-earth space that had not been investigated before. This more flexible approach (by the U.S. Air Force) should hopefully serve as an example to the National Aeronautics and Space Administration (NASA), where long preparation times and administrative burdens frustrate many scientists. The majority of NASA dollars for magnetospheric research in the next decade have been committed to the space shuttle program. This will be of little interest to auroral and magnetospheric physicists until polar orbits are flown, which will not be for some time. Use of the space shuttle as a platform for the inexpensive launching of smaller unmanned satellites may prove to be its most useful function.

It would be far-fetched to try to justify auroral research in terms of possible practical benefits to mankind. But certainly there are some direct effects of auroras and accompanying current systems that do affect our life on the earth's surface. High-frequency radio and television communications often show strong interference and even complete blackout during auroral displays, sometimes leading to loss of communications on commercial aircraft flights. Now that satellites have become important intercontinental communication links, the degradation of signal by increased ionospheric scintillations accompanying auroras has become an annoying problem. Even more serious is the spacecraft charging that takes place when high densities of auroral particles are injected into the inner magnetosphere during substorms; this effect has only recently been recognized as being responsible for complete failure of some multimillion dollar spacecraft, and future spacecraft design must guard against this hazard.

The military has a vital interest in the aurora, not only from the point of view of communications but also because of its effects on over-the-horizon early warning radar systems. Strong auroras give radar reflections and scatter that can readily mask genuine signals from approaching aircrafts or missiles, and for the duration of the aurora the United States is particularly vulnerable to attack from the north over the polar cap.

Strong auroras also induce large currents in long telephone lines, causing the loss of telephone circuits. More seriously, power transmission lines are similarly affected, and strong auroras have at times resulted in the tripping of protective circuit breakers, leaving large areas

(Opposite, top) Television composite picture showing the appearance of a barium-painted field line. Barium was explosively launched from a rocket above Kauai, Hawaii, and traced the magnetic field more than 7000 km to the South Pacific, where it is seen here penetrating the atmosphere (the bright upper picture at 7200 km from the launch point).
(Bottom) Artist's concept of space shuttle operations.
(1) The vehicle is launched from Kennedy Space Center, Florida. (2) The shuttle will be able to place satellites into Earth orbit, as well as conduct shuttle-based experiments. (3) A feature of the shuttle system is the ability to retrieve payloads for repair and maintenance. (4) The shuttle lands (a powerless glide) on a conventional runway at Kennedy Space Center.

The earth photographed in ultraviolet light. The left half of the picture is dayglow from the sunlit atmosphere. Two bands of enhanced airglow are seen about the equator on the nightside of the earth. The bright band around the South Pole is the southern auroral oval. The northern auroral oval is sunlit. This picture was taken from the moon during the *Apollo* mission.

of the northern United States and Canada without power.

Oil companies have recently become aware of the aurora, as it is expected that aurorally induced electric fields will increase the rate of ion migration to new arctic pipelines and so accelerate the corrosion effects that could lead to pipeline rupture. And in the past few years the possible effects of auroral processes on low-altitude meteorology have again become a serious field of study.

Another not impossible practical application of auroral and magnetospheric research may result from the similarity between the motion of energetic charged particles in the magnetosphere and particle motion in the laboratory "mirror machines" that have been developed in attempts to control thermonuclear fusion. Success in this venture might solve energy problems in our increasingly energy-hungry society, at least for the foreseeable future.

And as technology takes us far above the earth's surface, we must be on our guard against personal contact with energetic magnetospheric particles. An astronaut in the wrong orbit in the magnetosphere could be killed in a matter of days by radiation overdose. Similarly, passengers on a supersonic airliner flying through the polar cap at the height of a large magnetic storm might find their future family plans drastically affected.

Of cosmological interest is the fact that the earth's complex plasma environment may be one of the simpler of the astrophysical systems on our universe. The existence, location, and colors of auroras on other planets would depend on the strength of the magnetic field and the density and composition of the atmosphere. Mercury has a relatively strong magnetic field but little atmosphere, Venus has a dense atmosphere but a weak magnetic field, and Mars has little in the way of either atmosphere or magnetic field, whereas Jupiter has both a dense atmosphere and a strong magnetic field.

The magnetosphere of Mercury has been investigated by the *Mariner* 10 spacecraft, which led to the prediction that auroras should exist between 50° and 57° of latitude on the dayside and between 25° and 35° on the nightside. The thin helium atmosphere of Mercury would result in auroras being very near the planet's surface and being visible in the characteristic spectral emissions of helium.

On Venus it is the dense ionosphere rather than the (weak) magnetic field which presents an obstacle to the flow of the solar wind. Particle precipitation probably occurs over large areas of both the day and the night atmosphere, giving widespread diffuse auroras over much of the planet. The light would be the familiar oxygen and nitrogen emissions of auroras on earth, with the addition of strong carbon monoxide and carbon dioxide emissions. The recent *Pioneer* spacecraft missions (1978) detected such carbon monoxide emissions on the nightside of the planet, and these were interpreted as being caused by particle precipitation.

The weak magnetic field of Mars combined with the very thin carbon dioxide atmosphere leads one to expect little in the way of auroral phenomena above the red planet. The *Mariner* and *Viking* spacecraft have not detected any auroral effects in Mar's atmosphere, though the possibility of a weak diffuse widespread glow cannot be discounted.

The giant planet Jupiter, with its dense atmosphere and strong magnetic field, may well host the most spectacular auroras of the solar system. Jupiter's atmosphere is primarily hydrogen and helium, with some methane and ammonia, so the aurora would be expected to be seen in the spectral emissions of these gases. The magnetosphere of Jupiter is some 10 times larger than that of the earth, while close to the planet the magnetic field is complex in configuration because of complex circulation patterns in the planet's interior. Thus Jupiter's nightside is expected to exhibit fantastic shapes of shimmering auroras, and the recent *Voyager* mission has supported this expectation. Auroral effects were seen (in ultraviolet light) even in the daylight atmosphere as

Aurora photographed over the Alaska pipeline. Electric fields associated with the aurora may significantly increase corrosion of the pipe.

This *Voyager I* image was taken of Jupiter's darkside on March 5, 1979. The exposure was 3.2 minutes, from a distance of 500,000 km. The long bright double streak is a 30,000-km span of a Jovian aurora, seen along the limb of the planet. The band reaches to 84°N. Jupiter's north pole is on the limb toward upper center. Later pictures showed more diffuse aurora extending down to at least 70°N. The other bright spots lower in the picture are probably lightning flashes.

Voyager approached the planet. As the spacecraft looped around the dark side of Jupiter, on March 5, 1979, an aurora was photographed around the planet's north pole.

Comprehension of the physical processes in our own magnetosphere, of which the aurora is but a part, will undoubtedly give useful insight into processes into other astrophysical objects. The substorm mechanism that converts accumulated magnetic field energy in large portions of space to kinetic energy of a small fraction of particles seems to be a fundamental cosmic mechanism. Similar processes are thought to operate in solar flares and could be important in radio and X ray stars, pulsars (225), and the giant magnetospheres of the galaxies.

These practical effects associated with the aurora are not offered as justification for spending future tax dollars on auroral and magnetospheric research. The only justification needed for that is man's obligation to himself to pursue his own curiosity wherever it may lead him. It is fortunate that the pure scientific research propagated by this curiosity has led to most of the technical innovations of our civilization, and this happenstance should mollify those who ask for applicability and societal relevance of research.

For anyone with an inquiring mind who has ever witnessed a beautiful auroral display, the only justification needed for continued study of the aurora is the aurora itself.

The aurora in a science-oriented cartoon strip.

18
Auroral Photography

Such evanescent figures, almost
impossible to describe in words, are
even more difficult to suggest with any
approach of truth, in pictures.

Edward A. Wilson
Album of Photographs and Sketches,
National Antarctic Expedition, 1903

Fig. 2. Nordlichtdraperie am 1. Februar 1892.

Fig. 1. Nordlichtdraperie am 5. Januar 1892.

The first successful photographs of the aurora, taken by Brendel in 1892.

The quotation from E. A. Wilson on the title page of this chapter expresses the frustration of painters and photographers with a feeling of never being able to do justice to the beauty of the aurora. At another time, Wilson wrote:

> Unhappily it is an impossible thing even to suggest in picture, for as the curtain appears to fold in one direction, it is moved out of sight in another, while the varying intensity of the vertical beams of light which comprise it, now brilliant, now vanishing altogether, now stealthily appearing or disappearing imperceptibly, gives the onlooker a strange feeling of expectation and bewilderment, to which is added the conviction that the whole is very beautiful, but quite impossible to represent on paper.

Seventy-five years later, modern photographers with the advantages of high-speed color films and ultrafast lenses still experience the same dissatisfaction with their auroral photographs. Even though the pictures published here in Chapter 1 are beautiful and spectacular, they can never match the subtle beauty of the living aurora. Recently, more success has been achieved with cinematography, though the rapid and subtle color changes and structure changes that one can see with the human eye are still impossible to capture on film.

In this chapter we recount the history of the development of auroral photography and give practical hints for the modern photographer.

The First Auroral Photograph

The last 25 years of the nineteenth century saw the rapid development of the technology of photography. Many attempts were made before the First Polar Year, 1882-1883, to photograph the aurora, but all were unsuccessful. In his article on aurora in the journal *Heaven and Earth,* published in 1889, Weinstein (492) presented two auroral drawings which he claimed were based on photographs taken by Sophus Tromholt while Tromholt was at Kautokeino in Norwegian Lapland during the first Polar Year. This statement is in error, as Baschin directly questioned Tromholt about it and was told that attempts at auroral photography were unsuccessful. Tromholt used the fastest photographic plates available and exposure times of 4–7 minutes and obtained no trace of an image.

Tromholt was, however, successful some years later. In a footnote in his book and in a *Nature* article entitled "Aurora at Christiania" (volume 31, page 477, 1885), he reported that on March 15, 1885, he obtained an auroral photograph with an exposure of 8½ minutes. As such long exposures result in blurred images because of auroral motions, they are of no scientific value, and Tromholt did not publish this photograph.

Herr Baschin traveled to Bossekop, Finland, in 1891-1892 to study auroral effects on magnetic fields. He was accompanied by Herr Martin Brendel, whose charge was to try and photograph the aurora. Brendel borrowed a camera that belonged to O. Jesse of Berlin, which had previously been used to photograph noctiluscent clouds. It had a 210-mm focal length lens, 60-mm diameter (f3.5), and a field of view of 20° × 30° imaged on 9 × 12 cm plates. High-sensitivity plates were obtained from the Schleussner'sche company, and others were sensitized by Brendel himself by wetting them in a bath of erythrosinsililberbad (a silver compound).

Their efforts met with success, and they were surprised to find that a 3-minute exposure gave an overexposed image on the plates. Exposures

Martin Brendel 1862-1939
German physicist

This is the 'photograph of the author' that Tromholt provides for his book *Under the Rays of the Aurora Borealis.*

were gradually reduced, and it was found that good images were often obtained with 7-second exposures. Two of these photographs are shown on page 252. The bottom photograph shows an auroral drapery photographed with 1-minute exposure on January 5, 1892, at 9:05 P.M., and the top photograph shows part of an auroral arc photographed with 7-second exposure on February 1, 1892, at 10:15 P.M.

These are apparently the first published auroral photographs, though they were not published until 1900 when they appeared in a German meteorological journal (40). Presumably, this delay in publication was because the photographs had no particular scientific significance. Perhaps their eventual publication was prompted by others beginning to have some success at auroral photography, so it seemed prudent to Baschin and Brendel to establish their priority.

Early Attempts at Color Photography

Störmer states (445, 447) that efforts were made in Norway from 1940 onward to take pictures of the aurora in color. The fastest films from Kodak and Agfa were tested, but satisfactory results were not obtained

254

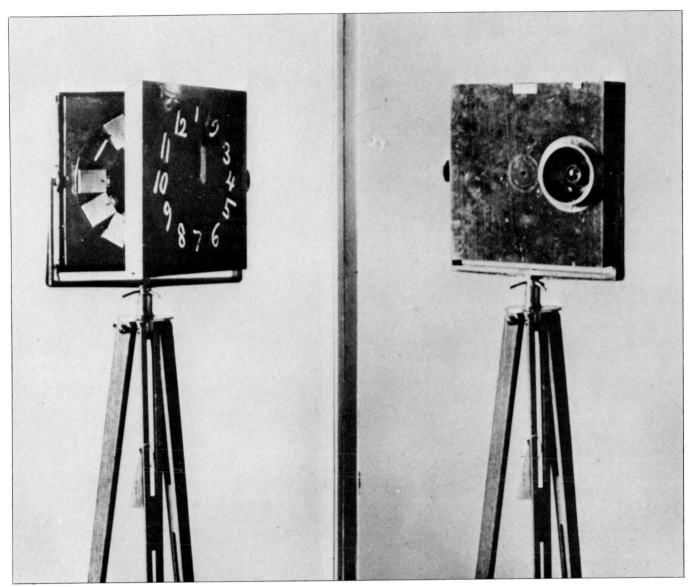

Störmer's auroral camera with clock timer.

except for quiet arcs and bands that allowed long exposure times (minutes).

Some of the first color auroral photographs in general circulation seem to have been shown by Gartlein (Cornell University) and Petrie (Saskatoon) at the Auroral Conference in London, Ontario, Canada, in 1951.

Life magazine published color pictures of the aurora in the June 8, 1953, issue, and since then countless color pictures of auroras have appeared in popular magazines, books, and scientific papers. Those published in this book represent a selection of some of the best pictures of the aurora obtained in the last 20 years.

Auroral Cinematography

Carl Störmer began experimenting in auroral photography (445, 447) in Norway in 1909. Fast lenses were desirable even at the expense of sharpness, so he searched photographic shops in Oslo and finally selected an f2.0 lens made by Ernemann of Dresden for a children's cine-camera. He successfully used this lens on his expedition to Bossekop in 1910, with exposures of 10 seconds to 1 minute.

255

In 1913 he attempted the first moving pictures of an aurora with this lens and the fastest film then available. The best results were obtained at midnight April 8, 1913, when a short series of pictures, each a 4-second exposure, was obtained. This was about 100 times the exposure time used in cine-cameras at that time, so when projected, the auroral movements would be speeded up by the same factor.

The next serious attempt at auroral cinematography was by Herr Bauer of Danzig Technical University (447), who was a guest worker at the Trömsø Observatory in Norway in 1931–1932. With a better lens and specially sensitized film, he succeeded in filming moderately bright auroras with exposure times of ¼–½ second. In 1938, Störmer used an f1.25 lens with Isopan ultrafilm (Agfa), and succeeded in taking film strips with exposures of about ⅓ second. When such films are projected, one gets a good impression of auroral development, but motions are speeded up by a factor of 5–10.

With modern high-speed lenses (f1.0) and high-speed black and white films (ASA ~2000) it is possible to photograph moderate auroras with exposures of about 1/10 second, so that motions are only speeded up by a factor of 3 when the film is projected.

It would now be possible to film even weak auroras in "real time" with modern electronic assistance of image intensifier cameras, though such techniques have not been applied to auroral cinematography.

Color cinematography has been more difficult because color films have been much less sensitive than black and white films. But high-speed color films available in the 1970's allowed this author to successfully film the aurora in color in 1973 with exposures of ½ second. Laboratory techniques of multiple-frame printing resulted in the auroral motions being speeded up by only a factor of 5 when the film was projected. Even faster color films were introduced in 1977; this film should allow exposures to be reduced by another factor of 2–3. It would also be possible to use image intensifier techniques combined with rotating colored filter wheels to achieve real time color cinematography, but such a project would be expensive and not warranted on scientific grounds.

Low light level television systems have been used over the last decade to record the aurora and have adequate sensitivity to capture even weak auroras in real time. More recently, a number of bore-sighted cameras fitted with colored filters have been used to obtain real time color television pictures. But the poor resolution, unreal colors, high contrast, and limited dynamic range of such systems do not give as aesthetically pleasing a result as regular color film.

Present photographic technology can come very close to fully capturing the beauty and ever changing nature of the aurora. Such films are not more readily available simply because of the costs involved in equipment, processing, and travel to arctic regions and the lack of demand for such film by popular media outlets.

Photographing the Aurora—Some Practical Hints

Although it is bright and visually spectacular to the dark-adapted human eye, the aurora is by no means as impressive to a fast color film. All auroral photographs require time exposure, the length of exposure depending on the auroral strength, the speed of the lens, and the ASA rating of the film used. During the time required to obtain satisfactory exposure, auroral forms will usually undergo some movement and cause some blurring of the image. (Note too that for time exposures, film speeds are no longer "linear" so that doubling the exposure does not have the same relative effect that it does for shorter exposure times, i.e., the "reciprocity effect.")

Auroral brightness is commonly classified in terms of the so-called International Brightness Coefficient (IBC), defined as follows:

International Brightness Coefficient	Description of Corresponding Auroral Intensity
0	Subvisual
1	Comparable with Milky Way
2	Comparable with moonlit cirrus clouds
3	Comparable with brightly lit cirrus or moonlit cumulus clouds
4	Much brighter than 3; casts discernible shadows

Some good rules of thumb are the following:

1. If any color is distinguishable in the aurora, then it must be at least an IBC 2 aurora. (Below the color threshold of the eye, IBC 1, the aurora appears to be white.)

2. If bright stars can no longer be seen through the aurora, it is probably IBC 3 or greater.

Many years of experience photographing the aurora suggest that the best results are obtained with Kodak's High Speed Ektachrome film (daylight type) with a nominal ASA rating of 160. Other brands of color films with a higher ASA rating are sold but are found to give less pleasing color and unrealistic reddish brown skies (instead of black) in parts of the picture where there is no aurora. High Speed Ektachrome film also behaves very well when "pushed" to higher speeds by special development. Tests have shown that development to speeds of ASA 640 (with nominal ASA 160 film) do not noticeably affect color balance, and the increase in film "graininess" is acceptable. Most large cities have processing laboratories offering this service at a slight extra charge. Kodak has recently (in the summer of 1978) introduced an even higher-speed color film, rated at ASA 400; they suggest that the film may be routinely pushed to ASA 800 and will give pleasing results even at ASA 1600. First tests indicate a color shift to the yellow with this forced processing, but this author has yet to test the film for auroral photography. If indeed it can be used at ASA 1600, the exposure times quoted below (at ASA 400) may be reduced by a further factor of 4.

The following tabulation gives recommended exposure times for a camera fitted with an f1.4 lens, with the assumption of a film speed of ASA 400.

Auroral Brightness	Recommended Exposure, seconds
IBC 1	20–30
IBC 2	10
IBC 3	1–2
IBC 4	¼

For different lens speeds these times should be multiplied by the factors in the following tabulation.

Exposure Adjustment Factors

Film Speed, ASA	Lens Speed f Number				
	1.0	1.4	2.0	2.8	4.0
50	4	8			
100	2	4	8		
200	1	2	4	8	
400	½	1	2	4	8
800	¼	½	1	2	4
1600	1/8	¼	½	1	2

Lens speeds of 2.8 and 4.0 and film speeds of 50 and 100 ASA are not recommended.

Some final tips:

1. Always use a tripod and shutter release cable.
2. Auroral photographs taken with a moon exceeding first quarter (half of disc illuminated) will be "washed out" and show little contrast.
3. Always try to include a foreground—houses, trees, etc. Wider angle shots are usually best.
4. If you are using your camera outside when it is very cold (less than, say, −10°C), always advance film very slowly to minimize the possibility of film cracking and static electricity. Some modern cameras with electronic shutters will not function properly at low temperatures.
5. It may be useful to note time and approximate direction for each photograph for later comparison with other photographers.

Good Luck!

19
The Aurora and Me

Felix qui potuit rerum cognoscere causas.
(Happy is he who gets to know the reason for things.)

Virgil (70–19 B.C.)
Georgics 2, 490
(Motto of Churchill College, Cambridge.
Local translation: It's great to
know what makes things tick.)

Research means going out into the unknown with
hopes of finding something new to bring home.

Albert Szent-Gyorgi
Perspectives in Biology and Medicine, 1971

My personal commitment to the aurora began by chance in New-castle, Australia, one Saturday morning in October 1961. I was soon to finish my undergraduate degree in physics and with my good friend John Hughes (who was finishing a degree in commerce) was scanning the *Sydney Morning Herald* looking for possible jobs. I do not recall who spoke first, but almost simultaneously we came across jobs which we thought suited the other. I pointed out an IBM advertisement to John (and he has since made his career in computers), and he showed me an advertisement placed by the Antarctic Division, Department of External Affairs, for physicists to spend one year at Australian bases in the Antarctic.

At my interview in Sydney about 6 weeks later, the Antarctic Division representative was Fred Jacka. On discovering that I had worked in undergraduate research on spectroscopy, Fred told me that I would be the auroral physicist at Mawson (67.6°S, 62.9°E) for 1963. Somehow I had managed to complete a physics course without ever learning just what an aurora was, so after accepting the job I headed directly for the nearby Mitchell Library in Sydney where I first read about the phenomenon that was to take me to all corners of the globe over the next 15 years.

The year 1962 was spent at the Antarctic Division in Melbourne preparing for the 1963 expedition. Apart from extensive reading in the field of auroral physics, I spent the time building and testing a new instrument designed and conceived by Fred Jacka—a tilting filter photometer. Technological advances at this time had produced the first really narrow optical interference filters, so that by tilting them a few degrees in the optic path it was possible to scan in wavelength. This would allow detection of very weak auroral emissions in the presence of background contamination. Fred's idea was to make real time measurements of the weak hydrogen (H_β) line in the aurora which results from protons bombarding the atmosphere. Although it was apparent at this time that auroras were mainly excited by precipitating electrons, the role of protons and their relationship to the electrons was still not clear.

Consequently, we had ordered one of these filters at H_β wavelength from Thin Film Products in Cambridge, Massachusetts. Thin Film Products was owned by Ed Barr, an artisan in the field of narrow filters. Ed's ability to make narrow filters is known worldwide, as is his seeming inability to answer correspondence or to make deliveries on schedule. As the sailing date for Antarctica approached, we became more and more concerned about receiving this one filter on which the whole year's research program depended. Letters progressed to cables, cable to international phone calls, and finally the filter arrived by diplomatic pouch the day before we sailed. (It was not to be the last time I received filters from Ed Barr at the last minute.)

On January 6, 1963, we sailed from Melbourne on the ice breaker *Nella Dan* for Antarctica. Four days out in the Great Australian Bight, our electronics technician developed acute appendicitis, and we had to put back into Albany, West Australia. Twenty-year old Bob White, a student of mine while I was tutoring at Queen's College, University of Melbourne, was flown across to join the ship as a last minute replacement. Also on board was a group of mountaineers led by Englishman Warrick Deacock; we left them on uninhabited Heard Island in the southern Indian Ocean, where they would (unsuccessfully) attempt to climb Big Ben, a precipitous mountain rising directly out of the sea to a snow capped 9,500 feet. (They returned two years later and successfully scaled the peak.)

The trip to Mawson took 4 weeks, for most of which I was seasick; one particular blizzard had us pointed into the wind for 4 days, getting nowhere and rolling up to 55°. It was a relief finally to penetrate the antarctic sea ice and sail smoothly for the last few days into Mawson.

Mawson is located on one of the few rock outcrops on the antarctic coast that remain free of ice all year. The daily katabatic winds (gravity winds of cold air simply flowing "downhill" from the antarctic plateau) ensure that any snow is quickly blown out to sea. After a strenuous week of manhandling a year's worth of supplies off the *Nella Dan,* I was one of 26 men abruptly left at that isolated outpost with no expectation of visitors or receiving mail until the *Nella Dan* returned about 12 months later. Our nearest neighbors were at another Australian Base, Davis, 400 miles to the east and a Russian base at Molodezhnaya, some 500 miles west.

The first task was to construct a new auroral observatory from prefabricated components and to install all our new equipment and check it out. Ably assisted by our carpenter Jock and our electronic engineer Don Creighton, we set up and were "ready for business" some 6 weeks later, just as a few hours of darkness each night were beginning to encroach on the constant 24-hour light of the antarctic summer.

It was one night in March 1963 that I saw my first spectacular aurora. We had seen an occasional glow or brief burst of light in the twilight skies of the previous weeks, but this particular night it was quite dark when I came out of the auroral observatory near midnight and set off down the hill to the main station. I was fast acquiring the habit of always looking up at the sky when outside at night, and this night I was treated to a splendid aurora australis. The words have all been used before, and no words can properly describe the awe-inspiring beauty of these silent heavenly displays: the rapidly moving forms, fluctuating in color and intensity, constantly changing and never repeating, sweeping across the sky in seconds. One feels very small and insignificant and at the same time somehow special and privileged for being perhaps the only one to see this particular aurora. I have seen two or three hundred similar displays since that first one at Mawson, yet I never tire of watching and always feel my confidence in my scientific understanding of what is happening sadly shaken as the display finally fades.

Icebergs just off the coast from Mawson, Antarctica.

The winter of observations at Mawson proceeded smoothly enough, thanks to the capabilities of Don Creighton. I would typically be up all night carrying out the observational program and would wake Don each morning as I was on my way to bed and inform him of the equipment problems of the night. When I joined Don that evening over my typical "breakfast" of roast beef or Irish stew, he would invariably inform me that everything was fixed for the evening. This being out of phase with the rest of the world is a problem that all auroral physicists must get used to, and one soon learns to accept bacon and eggs for "dinner" just before going to bed for the day.

Winter recreation at Mawson consisted of films, billiards, table tennis, and beer brewing as interior pursuits. At one time we had a three hole "golf course" out on the frozen sea ice. We used black golf balls, and local rules allowed tees on the "fairways." A well-hit drive will bounce forever on the smooth frozen surface, and I am sure many traveled over a mile.

Midwinter's day, June 21, is traditionally an enormous feast and party. For 6 weeks beforehand I choreographed our glaciologist, ionospheric physicist, and carpenter for a presentation on midwinter's day entitled *Duck Pond*, loosely based on the better known *Swan Lake*. The performance met with rave reviews and was repeated a number of times by popular demand.

As spring approached and the weather warmed, outdoor activities became more enjoyable—a little skiing after a fresh snow, ice skating behind a vehicle on the sea ice, and short dog sled trips to keep the huskies in trim and ready for glaciology field trips. Bob White, the young student who joined us at the last minute and worked as my electronics technician, suffered a bad fall on one of these dog trips—he collapsed some hours later and died the next day, a week before his twenty-first birthday. Bob is buried under a rock cairn on a peninsula near Mawson.

The end of winter and long hours of darkness brought the end of the aurora observational period, so the remaining 4 or 5 months waiting for the *Nella Dan* were spent beginning to organize and analyze the results and exploring around the Mawson area. Spring brings penguins and seals

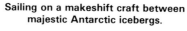

Sailing on a makeshift craft between majestic Antarctic icebergs.

and various birds, and I still fondly remember a young seal who adopted me, or vice versa; every day he would swim up and roll over in the water to be scratched under the chin.

We arrived back in Australia in March 1964 to that strange world of automobiles, women, and commerce; a world that had seen the assassination of President Kennedy and the arrival of the Beatles during our absence. I still remember 1963 for three things—Kennedy, the Beatles, and my first aurora.

I spent 1964 and 1965 in Sydney at the University of New South Wales, where I completed analysis of the Mawson results and wrote them up into my Ph.D. thesis. Essentially, we had clarified the role of protons in auroral phenomena and their basically independent behavior from the electrons. In addition, we completed a detailed study of how auroral electrons cause ionization that results in ionospheric absorption of radio waves.

So what does a fresh young Ph.D. with a specialty in auroral physics do in the real world? This was the question that confronted me in late 1965. It was impractical to continue working in this field in Australia because the only access to the aurora was via 12-month trips to Antarctica, and although I had fully enjoyed my antarctic sojourn, I had no desire for further long periods of isolation.

Thus it seemed the only solution was to try to find a job in the northern hemisphere, where the arctic and auroras were at least readily accessible by commercial aircraft. The only contact I had in the United States was Brian O'Brien, who had once worked for the Antarctic Division in Australia and then moved to the University of Iowa, where he worked with Van Allen's group during the exciting period of space exploration after the IGY. O'Brien had just moved to help start a new Space Science Department at Rice University in Houston, Texas. It did

Author with curious Weddell seal.

The author at Mawson, Antartica, 1963.

not seem to make much sense to be moving from Sydney, Australia, to Houston, Texas, to continue studies of the polar aurora but that was the reason that November 1965 found me on a slow boat across the Pacific to Mexico, thence by bus to Houston.

The Space Science Department at Rice University was bustling with activities in the mid-1960's—this was the height of NASA support of all activities to do with the upper atmosphere and space. O'Brien was deeply committed to a complex satellite program but was generous enough to allow me to follow my own interests, which consisted of building a new improved photometer and getting to the Arctic just as fast as I could.

Fort Churchill, on the western shore of Hudson's Bay in northern Canada, was the mecca for auroral scientists at this time. Complete launch facilities for rockets were available as well as observatory space for visiting scientists and appropriate technical and operational support. These facilities were jointly funded by the National Research Council of Canada and NASA. And it was possible to make commercial plane connections that took you from the humidity and 90° heat of Houston to the wind and −40°F cold of Fort Churchill in under 24 hours—quite an improvement on 4 weeks in a pitching ice-breaker!

Fort Churchill was also a very pleasant place to visit in those days; it was a dynamic little community of some 2,000. The local economy was based on support for the rocket range and the only hospital and school system for the Eskimos and Indians who lived in the hundreds of thousands of square miles of northern Manitoba. There were many young people among the teachers and hospital staff, and the two local clubs, appropriately the "Aurora Club" and the "Borealis Club," were welcome diversions when cloudy skies made auroral observations unprofitable.

Checking an auroral camera at dusk. The auroral observatory is at Ft. Churchill, Manitoba, Canada.

The scientific objective of my first trip to Fort Churchill in November 1966 was to determine accurately the efficiency of protons in exciting the main spectral lines in the aurora, so that in the future we could separate the effects of protons and electrons in the total excitation. At the same time, an unsuccessful search for helium emissions in the aurora put upper limits on the fluxes of "alpha particles" that might be accompanying the precipitating protons and electrons.

About this time Forrest Mozer from the University of California at Berkeley reported that his rocket experiments showed opposite behavior of protons and electrons, in that as the numbers of one increased, the other decreased, and vice versa. Such effects should then be evident in the auroral light generated, so I decided to accompany him on his next rocket campaign to Andoya, an island off the coast of northern Norway, and to provide photometric support of his experiment. This turned out to be my first experience of being entirely frustrated by the airlines; I spent a week twiddling my thumbs in Andoya while the airlines tried to locate my lost equipment. It was finally discovered in the back of a warehouse in Copenhagen, just in time to have it shipped directly back to Fort Churchill where I was to provide ground-based support for a rocket flight by one of our own graduate students from Rice.

One luxury of auroral physics is that it is seasonal. Observations are limited to the winter months, when there is sufficient darkness at polar latitudes. Consequently, one has the summer to pause and to reflect, to work on data, to plan new experiments, and to attend conferences to present new results. During the summer of 1967 I was fortunate enough to be able to present my new results on hydrogen and helium emissions in aurora at the Committee on Space Research conference in London and at a special conference honoring the memory of Kristian Birkeland at Sandefjord in Norway.

Brian O'Brien had been interested for some time in a special region off the coast of Brazil labeled the South Atlantic Anomaly. Here local anomalies in the earth's magnetic field bring regions of a given field strength much closer to the earth, and theorists had predicted that particles trapped in the earth's field would interact more deeply with the atmosphere here and precipitate, generating a weak low-latitude aurora in the process. Air Force measurements in the area had failed to detect auroral emissions and placed upper limits on the intensities, but with our techniques we knew we could measure lower intensities by a factor of 100 (if there was anything there to measure). The Office of Naval Research was sufficiently interested to sponsor the trip, so I set off from Rice with electronics engineer Foster Abney and headed for Rio de Janeiro to try to resolve the question. We had made contacts at the university in Rio, and they treated us royally, providing a hotel suite on the famous Copacabana beach (where I was able to indulge in surfing for the first time since leaving Sydney's beaches).

We obtained permission to set up our equipment at the site of a monument to the road builders who carved a highway through the rugged mountains west of Rio and elected to drive the 80 kilometers there and back every night rather than give up our accommodations on the beach. Two weeks of observations proved that if there was any auroral emission in the South Atlantic Anomaly, it was below the detection limit of even our equipment and hence was an unimportant effect.

O'Brien had plans to launch a rocket from Mar del Plata in Argentina

Preparation of rocket payload at Andoya, Norway, Forrest Mozer is on the left.

Rio de Janeiro, Brazil.

later that year to investigate particle precipitation in the anomaly. So he asked us to make some measurements from that region before we returned home. That request started a long comedy of errors which in retrospect is amusing but was far from enjoyable at the time.

We decided to fly to Montevideo in Uruguay and then to take the hydrofoil across the estuary of the Rio de la Plata to Buenos Aires. On landing in Montevideo I was told by their immigration authorities that my visa for Uruguay was out of date. I doubted this, as I had checked it just before leaving Houston, so we argued long and loud in our respective languages. They refused to let me enter the country and also decided that my Argentina visa was no good so that I could not even continue by plane to Argentina. Foster, being an American citizen, had no trouble, but for some reason Australians were not popular in Uruguay that year. I was allowed to call the British consul, who sportingly agreed to come out to the airport and "sign for me," but it would take him an hour to get there. I settled down to wait but 10 minutes later was suddenly dragged to my feet by two armed and uniformed gentlemen and escorted onto a plane leaving for Rio. Enroute we landed in Porto Alegre in southern Brazil, and having examined my passport in flight, I was aware that my Brazilian visa was good for just one entry, which I had already used up. Envisaging growing old on a constant shuttle between Rio and Montevideo, I decided that I might as well be hung for a sheep as a lamb and try to save some money on airfares. I secreted myself in the men's room at the airport in Porto Alegre until the plane left. On emerging I was immediately arrested, my passport was confiscated, and I was deposited at a small hotel with instructions not to leave.

The next morning brought some sanity when the police took me to the Argentinian consul who assured everyone that I had a perfectly valid visa; there were smiles and handshakes all around, and I was left to explore Porto Alegre until the first plane out to Buenos Aires that evening. I have since learned that situations like the one I experienced at Montevideo arise often in South America, are usually contrived, and can be quickly resolved with an appropriate bribe.

266

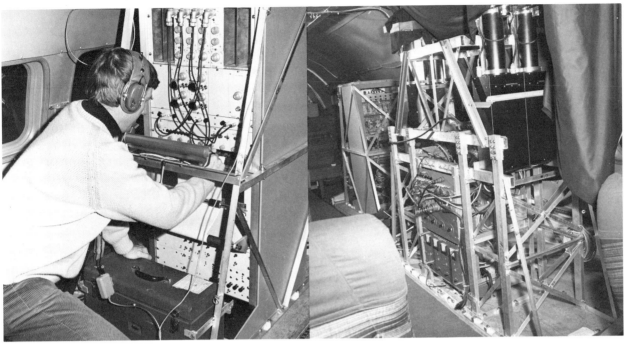

The author with photometric system installed on the NASA research aircraft *Galileo* for the 1967–1968 Airborne Auroral Expedition.

Buenos Aires was under water, in their worst flood in years. And the Argentinian customs people didn't like us either. With the help of the U.S. Embassy we spent a week advancing through a maze of officialdom and forms in quintriplicate until finally on a Friday afternoon we needed one more signature to have our equipment released from bond. But that signature belonged to a man who was trying to rescue his relatives in the flood and would not be back for over a week. We gave up and flew home. Our equipment arrived 6 months later. O'Brien never did launch that rocket from Argentina anyway.

NASA owned a Convair 990 plane, *Galileo,* which they used for various airborne research projects. In 1967 it was proposed to mount an Airborne Auroral Expedition which would fly out of Fort Churchill and conduct various latitudinal and longitudinal surveys of auroras. I went through the hassles of installing a new photometric system on the plane, which involved all sorts of engineering stress analyses of the structures that seemed to ensure that in the event of a crash the plane and its passengers might be scattered across the countryside, but the instruments would be standing in pristine tribute to the science of stress analysis. November 1967 saw groups from a dozen universities and institutions ready to take off from Ames Research Center in California for the shuttle trip to Fort Churchill. In all, we flew some 25 flights in the arctic region, and there are interesting anecdotes about a couple of them. One flight path took us on a latitudinal survey all the way to the north geomagnetic pole, which happens to be near Thule, Greenland, where we had to land to refuel. On board for the flight on January 21, 1968, was John Masterman from NBC news in Washington, whom the NASA Public Relations Office had invited along to do a story on the expedition. A couple of hours before we landed an international incident had developed at Thule. A Strategic Air Command B-52 bomber on airborne alert had crashed on the frozen ice of the harbor while coming in for an emergency landing. And somewhere out there either scattered across the ice or under several hundred feet of water were four hydrogen bombs! In the agreement with Denmark for landing rights at Thule, the

United States promised that no planes carrying nuclear weapons would overfly Greenland and Danish airspace, so the Danes were somewhat annoyed—irate might be a better term. There was a complete news blackout in force, and soon after we landed a ban was put on any unofficial outsiders coming in. But we had an NBC reporter with us, and no one at Thule knew about this; he was having the time of his life. Some hours later, as he was trying to interview higher and higher Air Force brass, the ignominious presence of the NASA Airborne Expedition was suddenly very embarrassing, and we were all roused from our beds and given 2 hours to get out of Greenland.

Of more scientific interest were our so-called "constant time flights." At these high latitudes, a jet plane flying west can keep at constant local time. The bright and spectacular aurora usually occur around midnight, and it was of interest to monitor midnight local time for a long period and to see how often these substorm aurora occurred and how long they lasted. We could take off from Fort Churchill at midnight, fly west for 5 or 6 hours and land at Fairbanks, Alaska, where it had just reached midnight. These flights showed that bright substorm aurora near the midnight meridian are intermittent, last perhaps 20 to 30 minutes, and repeat perhaps two or three times a day, depending on magnetic activity.

This was my first opportunity to visit College, Alaska, and the famous Geophysical Observatory of the University of Alaska. This institute has been responsible for much research on the aurora and has had many of the most respected scientists in the field on its staff, not the least of whom was Sydney Chapman, whom I was happy to meet for the first time.

The NASA Airborne Auroral Expedition was of great value, and one of the principal results was the discovery of a new zone of subvisual aurora generated by very low energy electrons precipitating poleward of the usual aurora. We presented these results at auroral conferences the following summers in Ås, Norway, and in Madrid.

Sydney Chapman was at the As conference, and I soon learned to appreciate his legendary physical condition. He was just 80 at that time, and each day we would go for a swim in a nearby lake. I would ride a bike the mile or so to the lake, while Chapman jogged alongside chatting. We would then swim across the lake and back, a distance of about a half mile, sometimes twice. Then I would ride back with Chapman still happily jogging alongside.

I left Rice University at this time and took an appointment at the Lockheed Palo Alto Research Laboratories. While at Rice I had become good friends with Stephen Mende. Stephen had left Hungary in 1956 and had been educated in England before coming to Rice to work on a television system for O'Brien's satellites. This task kept him occupied fully, and we never did collaborate on any projects while at Rice. He finally became disgruntled with the progress of the satellite program (that satellite was never launched) and moved to Lockheed. Some months later he called me looking for a graduate student to work with him. Although I had enjoyed 2½ good years at Rice, I was never particularly impressed by Houston and was getting itchy feet. I soon decided to move to Lockheed myself, and Stephen and I began a collaborative research effort that has continued ever since.

The following winter we took part in a second NASA Airborne Auroral Expedition, again based out of Fort Churchill, and essentially, we repeated earlier flight plans and accumulated better statistics. But we also completed two flights planned at the request of Syun-Ichi Akasofu of the University of Alaska to study the midday aurora. To do this, one must be in a location that is at the right latitude for aurora and is still dark at midday. In the northern hemisphere these conditions are satisfied for about a month in the middle of winter in a region out in the Arctic Ocean north of Norway. Consequently, we were based in Bodo, Norway, for a week and flew north to this unique region. Here we made the exciting discovery that there is a completely new type of aurora generated by very low energy electrons and protons—such particles excite an aurora (often subvisual)—that is almost monochromatically red in color. Because the particle energies involved were so low, we suggested that the solar wind might have direct access to this region through the "polar cusps" (see page 220). Satellite measurements published that same year showed that this was indeed the case.

A problem arose on our return to California from this trip. I had been in the United States for 4 years by then but had entered on a 2-year exchange visitors visa. This had duly expired and been renewed for a maximum of 1 year, so I had not had a valid visa for the previous 12 months. This is all right, provided you do not leave the country, as no one seems to keep too close a check on such things. On previous NASA flights, we had never been met by customs or immigration on our return, but on a recent research flight by another group to the Caribbean, various

The long and the short of auroral physics. Bill Fastie (Johns Hopkins University) and Syun Akasofu (University of Alaska) during the 1968–1969 Airborne Auroral Expedition.

Scientists taking part in the second Airborne Auroral Expedition in 1969 in Bodo, Norway.

Aurora photographed over a wing tip of *Galileo*.

(Opposite page) Pondering over an instrumental problem, South Pole, 1972. Ed Weber (right), Stephen Mende (seated), and the author.

passengers had been caught returning with more than their fair share of rum. I presume this is the reason why on our return from the Arctic we found customs and immigration waiting for us. With no valid visa, I was just a little dismayed by this. I explained my plight to my colleagues, and fortunately, the officials were operating from a makeshift desk in the hanger at Moffett Field. My friends crowded around the desk and I walked into the United States around the back of the group.

Another NASA airborne expedition to study the aurora and airglow was in the planning stages when disaster struck on April 12, 1973. The *Galileo* had been on a flight off the coast of California to test aerial camera techniques for detecting seals and walrus and was approaching the runway at Moffett Field when it was involved in a midair collision with a Navy P3 plane. All scientists and crew aboard, 13 men, were killed. Those of us who had flown the 45 aurora flights had come to know the pilots and crew well; we had come through a number of flight incidents in the harsh conditions of the Arctic, and it was a bitter irony that these men should then lose their lives in the clear and sunny skies of California.

Photometric measurements are limited to a given direction at a given time, so the next logical advance was to try to design an instrument to make measurements over the whole sky simultaneously, i.e., a photometric imaging system. Stephen had considerable experience and expertise with image intensifiers and television systems, and in 1969 he assembled a rather heavy and cumbersome system that consisted of a four-stage image intensifier and a television recording system. The image intensifier was water cooled, and we seemed to spend more time on the plumbing of that system than on the far more complex electronics. We spent the latter part of the 1969 winter in Alaska trying out the system and found that we could indeed record monochromatic (one wavelength) pictures of the aurora in real time for quite low auroral inten-

The author with photometers at Ft. Churchill, Manitoba, Canada.

sities. However, the optics of narrow bandwidth systems severely limit the field of view, so that we were only looking at 20° of the sky. It was clear after these experiments that a much larger field of view was desirable and that we would like to measure much lower intensities, even at the expense of time resolution.

In the summer of 1969, I left Lockheed and spent a pleasant 6 months in England, where I decided not to work and enjoyed a respite from the rather hectic research work of the previous 6 years. I played a lot of golf and squash, took advantage of the theatres and pubs of London, and arrived back in the United States refreshed to take up a new appointment at the Physics Department at Boston College. Even though we were now separated by the width of the continent, Stephen Mende and I continued our collaborative research on the aurora.

I decided to use the remaining months of that winter to return to northern Canada to carry out photometric measurements under the ATS 5 satellite. ATS 5 is a synchronous satellite which sits 22,000 miles from the earth's surface at a longitude of about 95°W. The magnetic field line passing through the satellite reaches the ground near Gillam, so that auroral measurements at this location can be compared to the satellite measurements of protons and electrons on the same field line. I was working alone on this trip, so after seeing some 10 crates of equipment through Canadian customs in Winnipeg and onto the first train north, I flew to Gillam and began looking for a suitable place to set up the equipment. At that time, Gillam was essentially a construction town for a vast hydroelectric dam project on the Churchill River. The Manitoba Hydro Authority was very cooperative in providing me with power and accommodations in their equipment storage yard, where I set up sensitive photometric and photographic equipment among the lumber, pipes, and vehicles for the dam project.

It soon becomes very lonely when one is working alone all night in an isolated spot, but in this case it was well worthwhile as both the aurora and the weather were cooperative and I knew I was obtaining some unique and important measurements. My most unpleasant recollection of the trip is the 1-mile walk every morning back into town, always into a 20–30 mile per hour wind with the temperature in the –40° to –50° range.

Because of our success at being able to detect the daytime polar cusp aurora photometrically and the growing interest in this region of the magnetosphere as the prime solar wind access area, Stephen and I decided to carry out a more comprehensive observational program of the region. Aside from the area over the Arctic Ocean, accessible only by airplane, the only other place that one can carry out observations of the midday aurora is from near the South Pole. Consequently, we obtained support from the Polar Programs Office of the National Science Foundation to set up equipment at South Pole Station, and January 1972 found Stephen and myself and Ed Weber, a graduate student from Boston College, bound for Antarctica via Hawaii, Fiji, and New Zealand. This was my first trip back since 1963, and this time a U.S. Navy Hercules aircraft took us from New Zealand to McMurdo Station on the antarctic coast in just 8 hours — a noisy and uncomfortable flight but preferable to the trip by ice breaker I had taken 9 years before.

The U.S. program in Antarctica is vast in comparison with those of other nations, with over 1,000 personnel on the continent during the summer months. Logistics support is provided by the U.S. Navy, with

Making film 'Spirits of the Polar Night';
interviewing Olav Devik in Oslo. Devik
was an assistant in Birkeland's
terella experiments.

funding from the National Science Foundation. Navy flight schedules in Antarctica seem contrived to assure that there are planes going everywhere except where you want to go. Consequently, it is not unusual to spend a month accomplishing a week's work. We did get our experiment set up at South Pole Station and waved good-bye to Ed, who was to remain there to operate the instruments through the long winter. During the 6 months of darkness the temperature falls from the balmy –30°F of summer down to below –100°F in winter. Ed spent 10 months there and obtained good data on midday auroras and on a rather spectacular solar flare event that occurred in August of that year. He subsequently used these data for his Ph.D. thesis at Boston College.

At about this time I became involved in making a documentary film about the aurora. I have always been a keen photographer and have taken many color slides of auroras over the years. But I have always wanted to try to capture this spectacular phenomena on color movie film. While I was at Mawson in 1963 I built my own all-sky time lapse camera from various optics spares and odds and ends we had around and obtained some encouraging results with a couple of seconds of exposure per frame. But my home-made camera did not frame accurately, so the image joggled when projected. I did not have the time or the finances to attack the problem again until 1972, when I obtained funding from Research Corporation to try to film the aurora, and additional funds from National Science Foundation to make the documentary film.

I succeeded in obtaining some spectacular color film of aurora from Fort Churchill in February 1973. It was still time lapse footage, but with various improvements and laboratory effects, the auroral motion was only speeded up by a factor of 5 when projected on the screen. In fact, this is ideal for presenting the phenomena, as you can show the spectacular substorm aurora in 3–4 minutes on the screen; the real time duration of 15–20 minutes would be too long to present as part of a film for the general public. To make the film, cinematographer David Westphal

273

and I traveled to Antarctica, Fort Churchill, Alaska, and Norway. The 30-minute film "Spirits of the Polar Night" was completed in late 1973 and combined history, folklore, and science in an interesting introduction to aurora for general audiences. We were pleased to win a number of prizes at international film festivals in the science documentary category.

Our scientific research continued and during the intervening summers, Stephen or I had reported our progress at various international conferences at Leningrad, Kingston (Canada), Dalseter (Norway), Moscow, Washington, and Kyoto (Japan). We felt we could now confidently interpret ground-based measurements of auroral spectra in terms of the type, flux, and energy of the precipitating auroral particles. The advantage of being able to do this from the ground is that you can study long-term changes, whereas rocket or polar satellites only make measurements for a few minutes or less as they pass above the aurora.

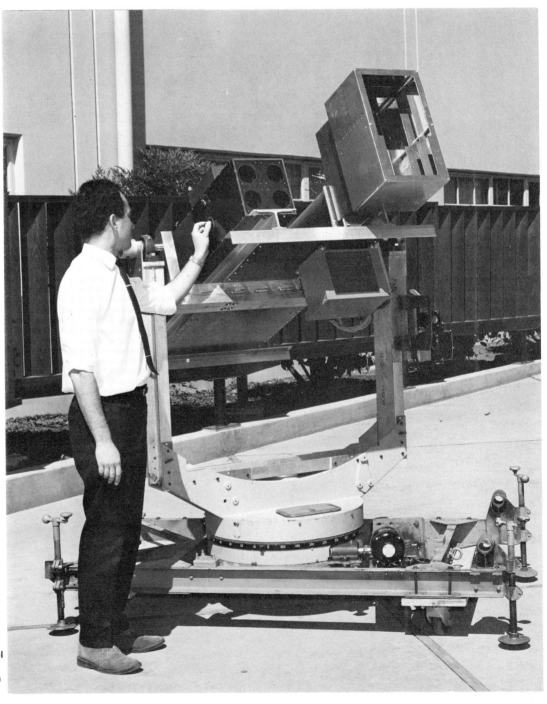

Stephen Mende with auroral instrumentation. On the left is a four-channel photometer and on the right an early version of a monochromatic television system (1969).

274

During 1972 we began to assemble more sophisticated equipment for all-sky monochromatic imaging, and in April 1973 we were in Fairbanks, Alaska, testing out the new system. At the same time we were maintaining our investigations in Antarctica, and Stephen returned to the south pole in December 1973 to move our photometers from South Pole Station to Siple Station, on the other side of the continent. We felt we had enough data on the midday aurora, whereas Siple was the scene of an exciting new experiment that could possibly produce artificial aurora. A group from Stanford University had constructed a 13-mile antenna across the ice to transmit radio waves up the magnetic field lines and into the magnetosphere. Interactions of these waves with charged particles around the earth often led to amplification of the waves before they were received at the other end of the field line in Roberval, Quebec. Theory predicted that these interactions should also lead to dumping of electrons into the atmosphere to cause induced aurora.

To cut a long story short, we found no evidence for this at Siple (nor at Roberval, where we conducted measurements for 3 weeks in November 1975); if there is an effect, it seemed that it was not common nor significant in terms of auroral light generated. Consequently, I returned to Antarctica and Siple Station in December 1974 to remove our equipment and send it back to the United States. Apart from the expected and frustrating logistics delays, the trip was uneventful until the day we left Siple. A Navy Hercules came in on skis and unloaded fuel, but with just three passengers on board, it could not lift off again from the soft snow. We spent over an hour taxiing up and down trying to compress a runway but all to no avail in lifting off. Finally, JATO (jet assisted take-off) cylinders were clamped to the fuselage, and we taxied out to try again. JATO's are always a little disturbing, as the cylinders have been known to break loose and damage the aircraft or to turn the aircraft nose over into the snow when they fire, and this was our crew's first experience with a JATO. But all went well. We got up to maximum speed, and the JATO's were fired, seeming to make little difference. The Hercules barely lifted off and crashed back down again, but it finally lifted again and stayed up to the cheers of the passengers, happy to be on the first leg of the long journey home.

In early 1975 we returned to Gillam again. A new synchronous satellite, ATS 6, had been launched with more sophisticated instrumentation to measure auroral particles. In addition to photometers, we were using our new and improved low-light television system that could photograph the complete sky at 1 wavelength and with very high sensitivity. Again we set up shop in a little hut in the Manitoba Hydro Authority storage yard and obtained excellent measurements over a 6-week period. A year later we carried out similar measurements from northern Sweden; ATS 6 had been moved to that location to transmit television programs to India, and fortunately, the magnetic field line through the satellite intersected the earth right near Kiruna, Sweden, where there is a fully equipped Geophysical Observatory. This time the aurora and weather did not cooperate, and only a few nights of good measurements were obtained during the 6-week observation period. Analysis of these data sets has shown a much more complex relationship than we had expected between the particles measured by the satellite and those traveling down the magnetic field lines to create the aurora.

I spent a lot of time in 1975 and early 1976 producing a new film, this

time about the magnetosphere. Almost 20 years had passed since the discovery of the Van Allen belts, and yet no film had been made to try to explain to the general public the exciting advances in our understanding of the space around the earth. It was a challenging film to make because you cannot go out and film the magnetosphere as you can film the aurora. The magnetosphere is a difficult abstract concept, so we decided on a historical-chronological approach to introduce the viewer gradually to more and more complex models, and yet we tried to keep the scientific content at a readily understood level. Funding for the film was provided by the National Science Foundation and NASA, and we shot on location throughout the United States, Europe, and the Arctic. The finished film, called "Earthspace," runs for 50 minutes and won a Golden Eagle Award at the Washington International Film Festival. We hope it will help inform the interested public as to just what scientists have learned in space over the past 20 years.

The year 1976 was the beginning of the 3-year International Magnetospheric Study. This is a similar though somewhat less ambitious project than the IGY and involves international cooperative efforts to extend our understanding of the magnetosphere and the aurora. To contribute to this effort, Stephen and I set up a three-station chain of photometers through central Canada. These instruments automatically scan the sky in a north-south direction and record the data on computer tape. When the weather cooperates and all stations have clear skies (15–20 nights each winter), we simultaneously obtain auroral data over 20° of latitude.

Recently, I have become interested in lower latitude phenomena. In late 1976 we installed a low-light all-sky imaging system on an Air Force research plane run by the Air Force Geophysics Laboratories. Besides flights to the auroral regions, there have been numerous expeditions to the equator over South America. The extreme sensitivity of our instrument allows time lapse films to be obtained of the very weak "airglow." These are very weak (subvisual) light emissions from the night sky which occur at all latitudes (see page 4). The airglow is usually quite uniform, but near the magnetic equator we found structured arcs of luminosity that move across the sky, sometimes with raylike structures reminiscent of the aurora. Indeed, when the time lapse films are projected, they look like the polar aurora! This equatorial phenomenon is not caused by particles bombarding the equatorial atmosphere but seems to result from gross spatial instabilities that affect the height of the ionosphere and hence the intensity of the airglow. We next plan to set up the instrument on the ground at Puerto Rico to do coordinated experiments with the giant radar dish at the Arecibo Observatory.

In 1977 we again sent a photometer to Siple Station and were successful in observing correlated auroral and wave effects, though these events were naturally occurring and not induced by the Siple transmitter. However, recent reports from Siple for the 1978 winter indicate that transmitter-induced auroral light pulses have finally been detected.

Stephen and I returned to Siple again in the 1979–1980 austral summer to install a low light level fisheye television system so that we can record the location, shape, and temporal characteristics of these man-made patches of auroral light.

I was fortunate to work in auroral physics for a 15-year period during which our understanding advanced more than it had in the previous 2,500 years. Not all the problems are solved (as was discussed in chapter 17), and future progress depends more and more on grander scientific endeavors. The days when an individual scientist could travel to polar regions hoping to make some significant contribution to the field seem to have passed. And with that passing, much of the character and satisfaction of auroral research is lost.

Increasingly, I have become involved in more long-term projects that promise rewards spread over longer periods of effort. For example, the chain of photometers that we operate in northern Canada took 2 years to construct and deploy. It will take another 2 years to collect a useful data base and probably 2 more years to analyze and compare these with other ground-based and satellite data. Stephen and I are experimenters on the first Space Shuttle research flights. The instrumentation will take 3 years to develop, the first flight being scheduled for 1981 and polar orbits not scheduled until 1983–1984. But probably the most significant future satellite experiments will not be carried out from the shuttle with its limiting low-altitude orbits. I hope to be involved with an exciting satellite imaging experiment that will photograph the whole global distribution of auroras over a period of many hours, enabling the first definitive studies of gross substorm morphology.

But whatever the future of auroral research and my part in it, I hope that the material in this book will convey something of the fascination of the aurora, and an appreciation of the life-long interest developed by anyone fortunate enough ever to witness a grand display.

Author at the geographic South Pole, January 1972. The polished dome was left to mark the South Pole by Paul Siple at the opening of the station at the beginning of the International Geophysical Year 1957.

278

Let the Spirits Play

What awesome sights are these Northern Lights, whatever be their meaning
As they dance and prance their ritual dance, shifting, fading, gleaming.
They pulse and beat and never repeat, show endless variation,
On a theme or a dream that always seems some mystic revelation.
Sibilant green with a vibrant sheen, purples, reds and yellow,
Muted shades that glow then fade, soft and rich and mellow.

Changing shapes from curtained drapes to dragons slowly creeping,
Curling tight then bursting bright like goats so nimbly leaping.
Flickering fires of funeral pyres of tortured souls departed,
Spirits at play at end of day, their nightly game just started.
A row of lances now advances, armies struggling in the sky,
Their proud parading slowly fading, disappearing from the eye.

What are these lights that fill the nights of arctic cold and silence
With flames and games of unknown names and hints of heavenly violence?
Do they foretell the funeral knell sending monarchs to damnation,
Or imminent wars bringing death to scores and endless devastation?
Is Mother Earth there giving birth from secret caverns distant,
To eerie lights that break long nights where the sun seems non-existent?

For centuries now men have wondered how this polar show is rendered,
A reflected glow from arctic snow or moonbeams caught suspended?
Perhaps the shroud of a cosmic cloud from far and unknown places,
Or the baleful glare from a comet's hair that swiftly onward races?
Is it a rent through Heaven's tent, where fire and brimstone's burning;
Or a glimpse through the door of Gods at war, lunging, twisting, turning?

Now men must know so men will go to the ends of the Earth to answer
The question why the arctic sky is reserved for the Merry Dancer.
They fight the snow and winds that blow through cold and endless nights,
To reach that place where face to face they meet those Northern Lights.
One day we'll look and find a book where mathematically treated
Will be the story of Nature's glory, examined, proved, completed!

But I'll not regret, or be upset, if I never see the day;
I prefer to dance in ignorance, and let the Spirits play!

R.H.E.

Appendix.   Names for the Aurora

The following table gives an idea of names that have been given to the aurora throughout the world and through the ages.

Names for the Auroras

Name	Literal Translation	Country or Region of Origin
chasmata	chasm	ancient Greece and Rome
hastae	javelin	ancient Greece and Rome
pithiae, tun	cask	ancient Greece and Rome
bothyneë	cave	ancient Greece and Rome
abysses	deep pit	ancient Greece and Rome
phasmata	sights	ancient Greece and Rome
trabes	beams	ancient Greece and Rome
bolides	darts	ancient Greece and Rome
faces	torch	ancient Greece and Rome
nubes	clouds	ancient Greece and Rome
sagitta	arrow	ancient Greece and Rome
pluvia sanguinea	blood rain	ancient Greece and Rome
pluvia lactea	milk rain	ancient Greece and Rome
sol noctu	night suns	ancient Greece and Rome
Chu-Long	candle dragon	ancient Chinese
thien lieh	cracks in heaven	Old Chinese
thien chiens	swords in heaven	Old Chinese
meteors, comets	general terms for heavenly phenomena, used for aurora	Europe up to the 19th century
capra saltans	dancing goat	16th century Europe
draco volans	flying dragon	16th century Europe
caelum ardens	fiery heavens	16th century Europe
acies igneae	battle lines of fire	16th century Europe
flying fires		16th century England
flying dragons		16th century England
fire drakes		16th century England
brennende Drachen	burning dragon	Old German
glühende Schlangen	glowing snakes	Old German
blutige Waffen	bloody weapons	Old German
feurige Spiesse	fiery spears	Old German
böse Geister	angry spirits	Old German
blutige Kriegsheere	bloody armies	Old German
Feuerzeichen	fire signs	Old German
Himmels Licht	heaven light	Old German
Nordschein	north shine	Old German
Nord Fluth	north stream	Old German
zarevnica	polar dawn	Old Russian
pazori	dawnlike	Old Russian
spoloxi	alarm or warning	Old Russian
spolochi chodjat	raging host passing	Siberia
Suratan-Tura	heaven god	Siberia
Looritz	name of a creature chained in a distant moor, of which aurora is a reflection—name used for aurora	Estonia
Virmalized taplevad	heavenly fight	Estonia
severnoe sijanie	northern radiance	Modern Russian
Bifrost	in Norse mythology, a bridge in the sky built by the Gods	Old Scandinavian

Name	Literal Translation	Country or Region of Origin
nordurljos	north light	13th century Norway
	Fire Lapp	Finland
vindlys	wind light	Finland
revontuli	fox fire	Finland
alme tulah	sky fires	Lapland
keoeeit	aurora shining during the period before there was any daylight	Eskimo
aksarnirq	ball player	Eskimo
sélamuit	sky dwellers	Eskimo
akshanik	that which moves rapidly	Eskimo
alugsukat	untimely birth	Eskimo
*Edthin**	caribou cow	North American Indian
	the old man	North American Indian
	dance of the dead	North American Indian
fir chlis	nimble men	Gaelic
merry dancers		Shetlands and Scotland
Noah's Ark†		Shetlands and Scotland
northern streamers		North England
burning spears		North England
Morris-dancers		North England
Lord Derwentwater's lights‡		North England
henbeams (henbanes) §		Hudson Bay area or North Scotland
pretty dancers		Canada and Scotland
marionettes		Canada
Buddha's lights		Ceylon
aurora borealis	northern dawn	Italy and France
aurora australis	southern dawn	France
aurora polaris	polar dawn	France
aurora septentrionalis	variation on aurora borealis	France
lucala boreale	variation on aurora borealis	France
lux borea	variation on aurora borealis	France
lumière Boréale	variation on aurora borealis	France
aurora tropicalis	suggested by S. Chapman for tropical auroras	
Nordlicht	northern light	Germany
Südlicht	southern light	Germany
Polarlicht	polar light	Germany
northern lights		England
southern lights		England
polar lights		England

*The Indians knew that when a hairy deer or caribou skin was briskly stroked with the hand on a dark night, it would emit many sparks of electrical fire, similar to the auroral light.

†Name given to multiple arcs across the zenith appearing to converge at two opposite points near the horizon.

‡Tradition has it that the aurora was seen on the night of February 23, 1716, the eve of the execution of the rebel lords Derwentwater and Kenmure. (They were beheaded for their part in the Jacobite uprising of 1715.) The great aurora seen by Halley was on March 6, 1716 (old system), and it is difficult to say whether there was another display 2 weeks earlier (two large displays in 2 weeks seems improbable), or whether the aurora of March 6 and the Earl's death became intermingled in the popular mind. The news of the death would probably have arrived in London around March 6, the delay being the usual travel time for the Newcastle to London coach.

§The origin of these names in not clear; ''henbeam'' may have been a type of square dance, whereas ''henbane'' is a poisonous plant.

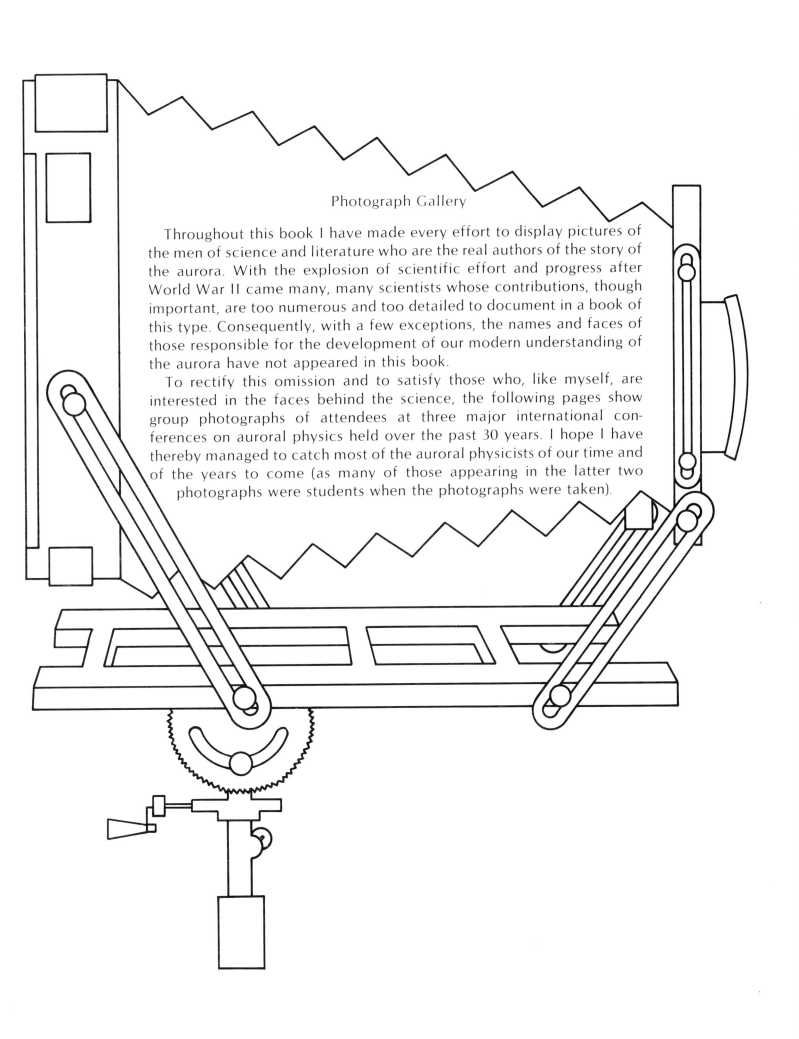

Photograph Gallery

Throughout this book I have made every effort to display pictures of the men of science and literature who are the real authors of the story of the aurora. With the explosion of scientific effort and progress after World War II came many, many scientists whose contributions, though important, are too numerous and too detailed to document in a book of this type. Consequently, with a few exceptions, the names and faces of those responsible for the development of our modern understanding of the aurora have not appeared in this book.

To rectify this omission and to satisfy those who, like myself, are interested in the faces behind the science, the following pages show group photographs of attendees at three major international conferences on auroral physics held over the past 30 years. I hope I have thereby managed to catch most of the auroral physicists of our time and of the years to come (as many of those appearing in the latter two photographs were students when the photographs were taken).

Conference on Auroral Physics   1951

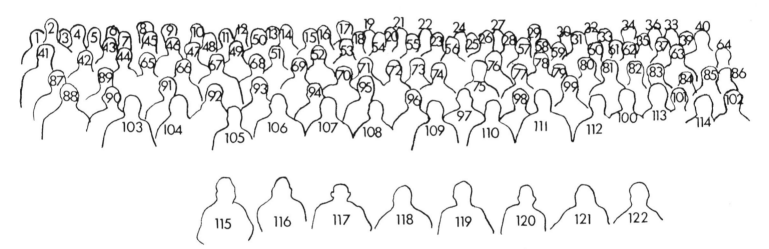

NATO Advanced Study Institute — Aurora and Airglow   1968

Summer Advanced Study Institute   1970

CONFERENCE ON AURORAL PHYSICS
UNIVERSITY OF WESTERN ONTARIO, LONDON,
ONTARIO, CANADA
JULY 23-26, 1951

1. H. E. Moses
   New York University
2. J. F. Carlson
   Iowa State College
3. A. T. Vassy
   Université de Paris, France
4. C. Störmer
   The University Observatory, Norway
5. M. E. Warga
   University of Pittsburgh
6. N. J. Oliver
   Air Force Cambridge Research Center
7. R. M. Chapman
   Air Force Cambridge Research Center
8. B. T. Darling
   Ohio State University
9. L. Herman
   Astrophysical Observatory, France
10. A. B. Meinel
    University of Chicago
11. T. Y. Wu
    National Research Council, Canada
12. S. Borowitz
    New York University

13. L. Katz
    Air Force Cambridge Research Center
14. N. C. Gerson
    Air Force Cambridge Research Center
15. D. Barbier
    Institut d'Astrophysique, France
16. R. W. Nicholls
    University of Western Ontario
17. D. M. Hunten
    University of Saskatchewan
18. S. Chapman
    Queen's College, England
19. H. S. W. Massey
    University College, England
20. C. W. Gartlein
    Cornell University
21. J. A. Vandertuin
    Air Force Cambridge Research Center
22. E. Vassy
    Université de Paris, France
23. W. Petrie
    University of Saskatchewan, Canada
24. R. G. Turner
    University of Western Ontario, Canada

25. D. Schulte
    University of Chicago
26. A. L. Aden
    Air Force Cambridge Research Center
27. H. Alfvén
    The Royal Institute of Technology, Sweden
28. A. D. Misener
    University of Western Ontario, Canada
29. D. R. Bates
    Queen's University of Belfast, Northern Ireland
30. R. W. B. Pearse
    Imperial College, England
31. M. W. Feast
    National Research Council, Canada
32. S. Altschuler
    Iowa State College
33. J. H. Blackwell
    University of Western Ontario, Canada
34. C. E. Montgomery
    University of Western Ontario, Canada
35. O. Oldenberg
    Harvard University

NATO ADVANCED STUDY INSTITUTE
AURORA AND AIRGLOW
ÅS, NORWAY
JULY 29 TO AUGUST 9, 1968

1. James C. Armstrong
   Johns Hopkins University
2. Newrick K. Reay
   Culham Laboratory, England
3. Harald Derblom
   Uppsala Ionospheric Observatory, Sweden
4. Keith Burrows
   SRC Radio and Space Research Station, England
5. G. Roger Pilkington
   University of Calgary, Canada
6. Roger Godard
   Groupe de Recherches Ionosphériques, France
7. Gunnar J. Kvifte
   Auroral Observatory, Norway
8. Wayne F. J. Evans
   University of Saskatchewan, Canada
9. M. W. J. Scourfield
   University of Calgary, Canada
10. T. Stockflet Jørgensen
    Det. Danske Meteorologiske Inst., Denmark
11. Peter Stauning
    Technical University of Denmark, Denmark
12. Thomas A. Clark
    University College, England
13. N. R. Parsons
    University of Calgary, Canada
14. Trygve R. Larsen
    Auroral Observatory, Norway
15. Rudolph Wiens
    University of Malawi, Malawi
16. Fokke Creutzberg
    National Research Council, Canada
17. Halvor R. Lindalen
    University of Bergen, Norway
18. Rolf Amundsen
    University of Bergen, Norway

19. Arne Haug
    Auroral Observatory, Norway
20. Einar Gjøen
    University of Bergen, Norway
21. Olav Holt
    Auroral Observatory, Norway
22. Helge Pettersen
    Auroral Observatory, Norway
23. A. W. Harrison
    University of Calgary, Canada
24. Michel Fehrenback
    Institut d'Astrophysique, France
25. James R. W. Hunter
    University of Sussex, England
26. Robert T. Brinkmann
    California Institute of Technology
27. Harald Schütz
    MPI Aeronomie, West Germany
28. William L. Imhof
    Lockheed Palo Alto Research Laboratory
29. Walter Landensperger
    Arbeitsgruppe für Physikalische, West Germany
30. David S. Evans
    Goddard Space Flight Center
31. Russel Vernon
    Bellcomm
32. Teodoro J. Vives
    Observatorio de Cartuja
33. Stein Ullaland
    University of Bergen, Norway
34. John Stadsnes
    University of Bergen, Norway
35. Eilert J. Naustvik
    Norwegian Institute of Cosmic Physics, Norway
36. R. L. Aggson
    Goddard Space Flight Center

37. Jacques Blamont
    CNRS Service d'Aéronomie, France
38. Kjeld Frellesvig
    Det. Danske Meteorologiske Inst., Denmark
39. Kjell Aarsnes
    University of Bergen, Norway
40. Karl Måseide
    Norwegian Institute of Cosmic Physics, Norway
41. Joachim Dachs
    Ruhr University, West Germany
42. Gordon F. Lyon
    University of Western Ontario, Canada
43. Marie-Lise Chanin
    CNRS Service d'Aéronomie, France
44. J. K. Hargreaves
    ESSA Research Laboratory
45. Dieter Beran
    DVL Inst. für Physik der Atmosphäre, West Germany
46. Gordon G. Shepherd
    University of Saskatchewan, Canada
47. Gunnar Skovli
    Norwegian Defence Research Est., Norway
48. Knud Lassen
    Det. Danske Meteorologiske Inst., Denmark
49. Albin Rossbach
    DVL Inst. für Physik der Atmosphäre, West Germany
50. Lars P. Block
    Royal Institute of Technology, Sweden
51. Robert Meier
    Naval Research Laboratory
52. Klaus Wilhelm
    MPI Aeronomie, West Germany
53. Melvin N. Oliven
    University of Iowa
54. Jorma Kangas
    University of Oulu, Finland

286

55. Jochen Münch
    MPI Aeronomie, West Germany
56. Robert O'Neil
    Air Force Cambridge Research Center
57. Lance Thomas
    SRC Radio and Space Research Station, England
58. Howard N. Rundle
    University of Saskatchewan, Canada
59. James R. Sudworth
    University of Exeter, England
60. Wesley A. Traub
    Smithsonian Astrophysical Observatory
61. Marcel Ackerman
    Institut d'Aéronomie Spatiale de Belgique, Belgium
62. Paul R. Satterblom
63. Andrew F. D. Scott
    University College, England
64. Ove Harang
    Auroral Observatory, Norway
65. Heinrich Raethjen
    Inst. für Kernphysik
66. Hans-Jürgen Braun
    MPI Aeronomie, West Germany
67. Günthur Lange-Hesse
    MPI Aeronomie, West Germany
68. William Swider
    Air Force Cambridge Research Center
69. Alan Moorwood
    University College, England
70. Kenneth I. Mayne
    University of Edinburgh, Scotland
71. Ero Kataja
    Geophysical Observatory of Finland, Finland
72. David F. Bleil
    U.S. Naval Ordnance Laboratory
73. Volker Zürn
    University of Göttingen, West Germany
74. Charles R. Chappell
    Huntingdon College
75. Rolf-Dieter Auer
    MPI Extraterrestrische Physik, West Germany
76. Eigil Hesstvedt
    University of Olso, Norway
77. Siegfried W. Drapatz
    MPI Extraterrestrische Physik, West Germany

78. Ernest Huppi
    Air Force Cambridge Research Center
79. Robert W. Carlson
    University of Southern California
80. Giovanni P. Gregori
    Istituto di Fisica dell'Atmosfera, Italy
81. Terrence S. Brown
    MPI Physik and Astrophysic, West Germany
82. Donald M. Hunten
    Kitt Peak National Observatory
83. Gisle Bjontegaard
    Norwegian Institute of Cosmic Physics, Norway
84. Marie-Louise Dubolin
    Laboratoire des Recherches Physiques, France
85. Kenneth C. Clark
    University of Washington
86. Jeannine Christophe
    Institut d'Astrophysique, France
87. Jean-Michel Bosqued
    Centre d'Etude Spatiale des Rayonnements, France
88. Sam Silverman
    Air Force Cambridge Research Center
89. Erika Preuschen
    MPI Aeronomie, West Germany
90. Richmond J. Hoch
    Battelle Memorial Institute
91. Allister Vallance Jones
    National Research Council, Canada
92. Charles S. Deehr
    University of Alaska
93. Robert Eather
    Lockheed Palo Alto Research Laboratory
94. Theodore W. Speiser
    University of Colorado
95. Alex E. S. Green
    University of Florida
96. L. D. de Feiter
    Ruksuniversiteit
97. Ian L. Thomas
    University of Melbourne, Australia
98. Roger W. Smith
    Queens University, Ireland
99. David C. Cartwright
    University of Colorado
100. Erich Rieger
    MPI Extraterrestrische Physik, West Germany

101. Derek J. Stone
    University of Exeter, England
102. Michael Isherwood
    University of Exeter, England
103. Apostolos Ch. Frangos
    Scientific Research Group on Space
104. Willy Stoffregen
    Uppsala Ionospheric Observatory, Sweden
105. Syun-I. Akasofu
    University of Alaska
106. Henry M. Morozumi
    General Electric Company
107. Brian A. Tinsley
    Southwest Center for Advanced Studies
108. Roger Pastiels
    Institut d'Aéronomie Spatiale de Belgique, Belgium
109. P. V. Kulkarni
    Physical Research Laboratory
110. Franco M. Vivona
    Istituto di Fisica dell'Atmosfera, Italy
111. Carlo Valenti
    Istituto di Fisica dell'Atmosfera, Italy
112. G. Martin Courtier
    SRC Radio and Space Research Station, England
113. Hans Lauche
    MPI Aeronomie, West Germany
114. Pablo Lagos
    Instituto Geofisica del Peru, Peru
115. Anders Omholt
    University of Oslo, Norway
116. Franklin E. Roach
    Boulder, Colorado
117. Martin Walt
    Lockheed Palo Alto Research Laboratory
118. Gilbert Weill
    Institut d'Astrophysique, France
119. Sydney Chapman
    University of Alaska
120. G. Kvifte
    Agricultural College of Norway, Norway
121. Leiv Harang
    University of Oslo, Norway
122. Billy M. McCormac
    Lockheed Palo Alto Research Laboratory

SUMMER ADVANCED STUDY INSTITUTE, AURORAL PHYSICS
QUEEN'S UNIVERSITY, KINGSTON, ONTARIO, CANADA
AUGUST 3-14, 1970

1. Samuel R. LaValle
   Aerospace Corporation
2. J. Ronald Burrows
   National Research Council, Canada
3. John R. Miller
   National Research Council, Canada
4. Gary Culp
   AFTAC
5. Melvin G. Heaps
   Utah State University
6. Kay D. Baker
   University of Utah
7. Vasant Agashe
   York University
8. Helge Pettersen
   The Auroral Observatory, Norway
9. Gerald Thuillier
10. Martin Walt
    Lockheed Palo Alto Research Laboratory

11. Catherine Vallance Jones
12. Alister Vallance Jones
    National Research Council, Canada
13. Billy M. McCormac
    Lockheed Palo Alto Research Laboratory
14. Dianne McCormac
15. Efstathios Kamaratos
    York University, Canada
16. L. Herman
    France
17. Hans Lauche
    Max-Planck-Institut für Aeronomie, West Germany
18. Michel Hamelin
    Groupe de Recherches Ionosphériques, France
19. David G. Nicholls
    University of Saskatchewan, Canada
20. W. F. J. Evans
    University of Saskatchewan, Canada

21. Guy Moreels
    CNRS, France
22. Mukhtar Ahmed
    Regis College
23. Gordon G. Shepherd
    York University, Canada
24. Richard G. Johnson
    Lockheed Palo Alto Research Laboratory
25. Margaret P. Heeran
    Holy Rosary Convent, Ireland
26. Doris H. Jelly
    Communications Research Centre, Canada
27. Jacques Vigneron
    Laboratoire de l'Observatorie, France
28. Jeannine Christophe
    Institut d'Astrophysique, France
29. A. C. Levasseur
    CNES, France

30. Dieter Beran
Institut für Physik der Atmosphere,
West Germany
31. Axel Korth
Max-Planck-Institut für Stratosphärenphysik,
West Germany
32. G. M. Shah
Meteorology Branch, Canada
33. Anthony Lui
34. Harry M. Sullivan
University of Victoria, Canada
35. James H. McElaney
Fairfield University
36. Giovanni O. Gregori
Istituto di Fisica Dell'Atmosfera, Italy
37. William Sawchuck
University of Alberta, Canada
38. Hugh C. Wood
University of Saskatchewan, Canada
39. Donald Hunten
Kitt Peak National Observatory
40. D. J. McEwen
University of Saskatchewan, Canada
41. Jean-Claude Gerard
Université de Liege, Belgium
42. Rainer Kist
Arbeitsgruppe für Physikalische,
West Germany
43. R. Potelette
Groupe de Recherches Ionosphériques,
France
44. Paul Simon
Institut d'Aéronomie Spatiale de Belgique,
Belgium
45. William R. Sheldon
University of Houston
46. Eric Hewstone Carman
U.B.L.S., South Africa
47. D. R. Snelling
Defense Research Establishment Valcartier,
Canada
48. Edward J. Llewellyn
University of Saskatchewan, Canada
49. Walter E. Brown
University of Hawaii
50. Alev Egeland
The Norwegian Institute of Cosmic Physics,
Norway
51. J. R. Sternberg
Oxford University, England
52. David J. Marsh
University of Southampton,
England
53. Brian A. Whalen
National Research Council, Canada

54. Lawrence D. Kavanagh, Jr.
NASA Headquarters
55. Stephen B. Mende
Lockheed Palo Alto Research Laboratory
56. David Winningham
University of Texas at Dallas
57. Lawrence W. Choy
University of New Hampshire
58. Charles R. Chappell
Lockheed Palo Alto Research Laboratory
59. Walter J. Williams
University of Denver
60. Günther Lange-Hesse,
Max-Planck-Institut für Aeronomie,
West Germany
61. David F. Bleil
U.S. Naval Ordnance Laboratory
62. Hans G. Mayr
Goddard Space Flight Center
63. Roger W. Smith
Queens University of Belfast, Northern
Ireland
64. H. N. Rundle
University of Saskatchewan, Canada
65. Michael C. Isherwood
Queen's University of Belfast, Northern
Ireland
66. J. E. Evans
Lockheed Palo Alto Research Laboratory
67. Eric V. Pemberton
Brandon University, Canada
68. Guido Visconti
Università dell'Aquila, Italy
69. Johannes Steyn
70. A. Lawrence Spitz
National Research Council, Canada
71. L. E. Montbriand
Communications Research Center, Canada
72. S. R. Pal
Meteorological Branch, Canada
73. John F. Noxon
Harvard University
74. Akira Hasegawa
Bell Telephone Laboratories
75. George Frederick Stuart
Physics and Engineering Laboratory, New
Zealand
76. E. C. Zipf
University of Pittsburgh
77. T. R. Hartz
Communications Research Center, Canada
78. Jeffrey B. Pearce
University of Colorado
79. Robert C. Schaeffer
Johns Hopkins University

80. Dan Holland
Lockheed Palo Alto Research Laboratory
81. Derek M. Cunnold
Massachusetts Institute of Technology
82. Rolf Boström
The Royal Institute of Technology, Sweden
83. James T. Coleman
Battelle Memorial Institute
84. Lance Thomas
SRC Radio and Space Research Station,
England
85. Adolph Johansson
86. Duncan A. Bryant
SRC Radio and Space Research Station,
England
87. R. L. Gattinger
National Research Council, Canada
88. J. N. Bradbury
Lockheed Palo Alto Research Laboratory
89. Thomas Hallinan
University of Alaska
90. Roland E. Johnson
University of Alaska
91. Kenneth D. McWatters
University of Michigan
92. Michael T. Schwitters
Illinois
93. F. E. Bunn
York University, Canada
94. Robert D. Sears
Lockheed Palo Alto Research Laboratory
95. David C. Cartwright
Aerospace Corporation
96. R. D. Rawcliffe
Aerospace Corporation
97. R. S. Unwin
Pel Auroral Station, New Zealand
98. A. G. McNamara
National Research Council, Canada
99. John W. Meriwether
University of Maryland
100. T. Neil Davis
University of Alaska
101. Charles R. Wilson
University of Alaska
102. Ian B. McDiarmid
National Research Council, Canada
103. Michael D. Watson
National Research Council, Canada
104. Robert H. Eather
Lockheed Palo Alto Research Laboratory
105. Luke J. Sennema
University of Western Ontario, Canada
106. Fokke Creutzberg
UAR/REED, Canada

Acknowledgments

Many friends and colleagues have offered helpful suggestions during the preparation of this book; in particular, I would like to thank Sam Silverman, Alv Egeland, and George Siscoe for drawing my attention to many interesting references.

The many people and organizations who generously allowed me to use their photographic material are acknowledged in the photography credits.

The historical aspects of this book were a pleasure to research, largely because of the excellent facilities of the Harvard University Libraries, the Yale University Libraries, the Boston Public Library, the Air Force Geophysics Laboratory Library, and the Scott Polar Institute Library. I would especially like to thank Sam Silverman at Air Force Geophysics Laboratory for allowing me access to his extensive collection of auroral material.

Heartfelt thanks are due M. Kathleen Caldwell (design) and Mary J. Scroggins (editor); both suffered cheerfully through many alterations and changes and a long series of "last" additions.

Finally, I express sincere appreciation to Joyce E. Vickery, who has typed a new "final" draft of this book three times over the last 3 years.

FIGURE ACKNOWLEDGMENTS

The author gratefully acknowledges permission to use the following material.

v: Courtesy, Fogg Art Museum, Harvard University.
viii: Bapst Library, Boston College.
Margin portraits are acknowledged separately, in order of appearance, within their corresponding chapter listings.

CHAPTER 1

page 4: (top) Courtesy, E. J. Weber, Air Force Geophysics Laboratory.
          (bottom) American Geophysical Union.
      7: Space Environment Services Center, Boulder, Colorado.
    8-9: 1   Air Force Geophysics Laboratory.
          2   Robert H. Eather.
          3   National Aeronautics and Space Administration. Photograph by Owen Garriott.
          4   Al McNeil, Calgary, Canada.
  10-13: 5-7 Robert H. Eather.
          8   Al Belon, Geophysical Institute, University of Alaska.
        9-26 Robert H. Eather.
     14: 27-28 George Cresswell, CSIRO, Cronulla, N.S.W., Australia.
         29   Robert H. Eather.
     15: 30   George Cresswell, CSIRO, Cronulla, N.S.W., Australia.
         31   Torbjörn Lövgren, Kiruna, Sweden.
  16-17: 32   Robert H. Eather.
  18-19: 33-34 Gustav Lamprecht, Clear, Alaska.
  20-21: 35-37 Robert H. Eather.
     22: 38, 40 Malcolm Lockwood, Fairbanks, Alaska.
         39   Phil Pazich, SAMSO, Los Angeles, California.
     23: 41   Malcolm Lockwood, Fairbanks, Alaska.
     24: 42   Victor Hessler, Boulder, Colorado.
         43   Gustav Lamprecht, Clear, Alaska.
     25: 44   Malcolm Lockwood, Fairbanks, Alaska.
     26: 45   George Cresswell, CSIRO, Cronulla, N.S.W., Australia.
         46   Lee Snyder, Air Force Geophysics Laboratory.
         47   Syun Akasofu, Geophysical Institute, University of Alaska.
       48-50 Al McNeil, Calgary, Canada.
     27: 51   Torbjörn Lövgren, Kiruna, Sweden.
       52-54 Al McNeil, Calgary, Canada.
       55-56 Torbjörn Lövgren, Kiruna, Sweden.
     28: Painting by Harald Moltke, copyright © The Arctic Institute, Danish Meteorological Institute.
     29: Courtesy, Gustav Lamprecht, Clear, Alaska.
  30-31: Air Force Geophysics Laboratory, arranged by Robert H. Eather.

CHAPTER 2

page 35: By permission of the Houghton Library, Harvard University.
      36: American Geophysical Union.
      37: Robert H. Eather.
      39: By permission of the Houghton Library, Harvard University.
      40: (bottom) Courtesy, P. K. Wang, University of California, Los Angeles.
      41: *Kexue Tangbao,* 20, 1975.
      43: (top, middle) Widener Library, Harvard University.
          (bottom) By permission of the Houghton Library, Harvard University.
      45: (top) Zentral Bibliothek, Zurich.
          (middle) Royal Observatory, Edinburgh.
          (bottom) Magyar Nemzeti Muzeum, Budapest.
      46: (top) Courtesy, Royal Library, Copenhagen.
          (bottom) Germanisches Nationalmuseum, Nürnberg.
Margin portraits: Aristotle: Courtesy, Burndy Library, Norwalk, Connecticut.
          Pliny: Reproduced from the collection of the Library of Congress.
          Lycosthenes: By permission of the Houghton Library, Harvard University.
          Seneca: Reproduced from the collection of the Library of Congress.
          Keimatsu: Kanazawa University, Japan.
          Schove: Courtesy, Justin Schove, St. David's College, Kent, England.
          Botley: Courtesy, C. M. Botley, Kent, England.
          Gregory of Tours: Librairie Larousse, Paris.
          Fulke: By permission of the Houghton Library, Harvard University.

CHAPTER 14

page   186:   Courtesy, Trustees of the Boston Public Library.
        189:    Widener Library, Harvard University.
        190:    Widener Library, Harvard University.
        191–194: Paintings by Harald Moltke, copyright © by The Arctic Institute, Danish Meteorological Institute.
        196:    Scott Polar Research Institute, Cambridge, England.
        201:    Widener Library, Harvard University.
        203:    Courtesy, G. P. Putnam's Sons, New York.
        204:    Painted by Stephen Hamilton, Amherst, Massachusetts. Courtesy, Robert H. Eather.
        205:    (Woodcut) Scott Polar Research Institute, Cambridge, England.
        207:    From J.R. Capron, *Aurora: Their Character and Spectra,* Spon, London, 1879, Air Force Geophysics Laboratory Library.
        208–209: Illustration copyright © 1977 by John Vernon Lord. Reproduced with permission.
        214:    National Aeronautics and Space Administration.
Margin portraits: Longfellow: Robert H. Eather library.
                Chaucer: Bapst Library, Boston College.
                Thomson: Bapst Library, Boston College.
                Collins: Bapst Library, Boston College.
                Smart: Bapst Library, Boston College.
                Coleridge: Bapst Library, Boston College.
                Milman: Bapst Library, Boston College.
                Goethe: Bapst Library, Boston College.
                Moltke: Copyright © The Arctic Institute, Danish Meteorological Institute.
                Byron: Robert H. Eather library.
                Keats: Robert H. Eather library.
                Tennyson: Robert H. Eather library.
                Scott: Robert H. Eather library.
                Burns: Robert H. Eather library.
                Southey: Bapst Library, Boston College.
                Browning: Robert H. Eather library.
                Cranch: Bapst Library, Boston College.
                Taylor: Widener Library, Harvard University.
                Aytoun: Bapst Library, Boston College.
                Whittier: Bapst Library, Boston College.
                Foran: Widener Library, Harvard University.
                Dickinson: Bapst Library, Boston College.
                Stevens: Bapst Library, Boston College.
                Weyprecht: Air Force Geophysics Laboratory Library.
                Nansen: Scott Polar Research Institute, Cambridge, England.
                Taylor: Bapst Library, Boston College.
                Byrd: Bapst Library, Boston College.
                Hopkins: Bapst Library, Boston College.
                Aiken: Bapst Library, Boston College.
                Service: Bapst Library, Boston College.

CHAPTER 15

page   216: National Aeronautics and Space Administration, collage by E. H. Rogers.
        218: (top) Courtesy, Hale Observatories, Pasadena, California.
             (bottom) Courtesy, *Nature.*
        219: (top) Kiruna Geophysics Institute Library, Sweden.
             (bottom) Courtesy, Gotz Paschman, Max Planck Institute, Garching.
        220: (top) Courtesy, J. G. Roederer, University of Alaska.
             (bottom left) Robert H. Eather.
             (bottom right) Courtesy, Y. Galperin, Space Research Institute, USSR Academy of Sciences.
        222: National Aeronautics and Space Administration.
        223: (top) Courtesy, L. Frank, University of Iowa.
             (bottom) Courtesy, C. McIlwain, University of California at San Diego.
        224: Courtesy, J. G. Roederer, University of Alaska.
        225: National Aeronautics and Space Administration.
        227: (top) National Aeronautics and Space Administration.
             (bottom) National Aeronautics and Space Administration, photograph by Owen Garriott.
        229: Pierre Mion, copyright © *National Geographic Magazine.*
        230: Courtesy, Walter Heikkila, University of Texas at Dallas.
Margin portraits: Gilbert: Bapst Library, Boston College.
                Gold: Photograph by John Stewart. Courtesy, T. Gold.
                Bierman: Courtesy, L. Bierman.
                Ferraro: Courtesy, Mrs. V. C. A. Ferraro.
                Alfvén: Courtesy, H. Alfvén.

CHAPTER 16:

page 232: Courtesy, Julianna Turkevich, Technology International Corporation and Defense Nuclear Agency.

234: From J. R. Capron, *Aurora: Their Character and Spectra,* Spon, London, 1879, Air Force Geophysics Laboratory Library.

235: (all) Widener Library, Harvard University.

236: Widener Library, Harvard University.

237: (top) Air Force Geophysics Laboratory Library.
(bottom) Composite from illustrations from *Space Science Reviews.*

238: National Aeronautics and Space Administration.

239: Courtesy, Walter Lang, Department of Physics and Astronomy, University of Hawaii.

240: (both) Courtesy, W. Hess, National Oceanographic and Atmospheric Administration, Boulder, Colorado.

242: Widener Library, Harvard University.

Margin portraits: Canton: Courtesy, National Portrait Gallery, London.

Capron: Courtesy, County Archivist, Guilford, Surrey, England.

Lemström: Courtesy, University of Helsinki, Finland.

Vaussenat: Courtesy, Observatoires du Pic-du-Midi et de Toulouse, France.

CHAPTER 17

page 244: Courtesy, National Oceanographic and Atmospheric Administration, Boulder, Colorado.

245: Robert H. Eather.

246: (top) Courtesy, E. M. Wescott, Geophysics Institute, University of Alaska.
(bottom) National Aeronautics and Space Administration.

248: Naval Research Laboratory, Washington, D. C.

249: Courtesy, R. G. Hochschild, Fairbanks, Alaska.

250: (top) Jet Propulsion Laboratory, California Institute of Technology, Pasadena.
(bottom) Excerpted from S. T. Butler and Robert Raymond, *The Family of the Sun: Frontiers of Science,* copyright © 1975 by Science Features. Reprinted by permission of Doubleday & Company, Inc.

CHAPTER 18

page 252: Air Force Geophysics Laboratory Library.

254: Widener Library, Harvard University.

255: Courtesy, Alv Egeland, University of Oslo.

Margin portrait: Brendel: Deutsche Bucherei, Leipzig.

CHAPTER 19

page 261: Robert H. Eather.

262: Robert H. Eather.

263: (both) Robert H. Eather.

264: Courtesy, David Westphal.

265: Robert H. Eather.

266: Robert H. Eather.

267: (both) National Aeronautics and Space Administration.

269: (top) Robert H. Eather.
(middle, bottom) National Aeronautics and Space Administration.

270: Robert H. Eather.

271: Courtesy, David Westphal.

273: Courtesy, David Westphal.

274: Lockheed Palo Alto Research Library.

276: Robert H. Eather.

278: Courtesy, David Westphal.

PICTURE GALLERY

1: Design by Deborah LeLait, American Geophysical Union.

2: Courtesy, Air Force Geophysics Laboratory Library.

3: Courtesy, Billy M. McCormac, Lockheed Palo Alto Research Laboratory.

4: Courtesy, Billy M. McCormac, Lockheed Palo Alto Research Laboratory.

Bibliography

References are arranged alphabetically and numbered consecutively. References are cited in text by smaller parenthetical numbers that correspond to the appropriate reference in this list. Not all references here are called out in the text, though all were consulted while writing this book. Omitted from the list are many hundreds of papers describing particular auroras, many auroral catalogs, and articles and books on scores of polar expeditions that discuss the aurora. There has been no attempt to reference more recent specialized scientific papers unless they are specifically referred to in the text.

References that are good historical sources, contain good reviews of their subject, or are good general reading are preceded by an asterisk (*).

There is a list of films about the aurora following the bibliography.

1. Abbe, C., The altitude of the aurora above the earth's surface, *Terr. Magn. Atmos. Elec., 3,* 5, 1898.
2. Adolph, W., *A New Theory of the Solar System,* p. 42, Catholic Publishing and Bookselling, London, 1859.
3. Aiken, C., Tigermine, in *Who's Zoo,* Atheneum, New York, 1977.
4. Akasofu, S.-I., The development of the auroral substorm, *Planet. Space Sci., 12,* 273, 1964.
5. *Akasofu, S.-I., The aurora, *Sci. Amer., 213,* 55, 1965.
6. Akasofu, S.-I. (Ed.), *Sydney Chapman Eighty, From His Friends,* University of Colorado Press, Boulder, 1969.
7. *Akasofu, S.-I., The aurora from ancient times, Geophys. Inst., Univ. of Alas., Annual Report, 1973–1974.
8. *Akasofu, S.-I., The aurora and the magnetosphere: The Chapman memorial lecture, *Planet. Space Sci., 22,* 885, 1974.
9. Akasofu, S.-I., A study of auroral displays photographed from the DMSP-2 satellite and the Alaska meridian chain of stations, *Space Sci. Rev., 16,* 617, 1974.
10. Akasofu, S.-I., The aurora: An electrical discharge processs around the earth, *Endeavor, 2* (1), 7, 1978.
11. *Akasofu, S.-I., *Aurora Borealis–The Amazing Northern Lights, Alaska Geographics,* vol. 6, Alaska, Geographic Society, Anchorage, Alas., 1979.
12. *Akasofu, S.-I., and L. J. Lanzerotti, The earth's magnetosphere, *Phys. Today, 28,* 28, 1975.
13. Akenside, M., Hymn to science, in *The Poetical Works of Mark Akenside,* edited by T. Park, Suttaby, Evance, and Hutchings, London, 1811.
14. Alfvén, H., *Cosmological Electrodynamics,* Oxford University Press, London, 1950.
15. Alfvén, H., Note on the 'auroral oval,' *J. Geophys. Res., 72,* 3503, 1967.
16. Amanuensis (pen name), *Mathematical, Geometrical and Philosophical Delights,* vol. 1, edited by T. Whiting, London, 1792.
17. Ames, N., *Nathaniel Ames' Almanack,* J. Draper, Boston, Mass., 1731.
18. Amundsen, R., *The South Pole,* translated from Norwegian by A. G. Chater, J. Murray, London, 1912.
19. Angenheister, G., and C. J. Westland, The magnetic storm of May 13–16, 1921 at Apia, Samoa, and the aurora borealis, *Terr. Magn. Atmos. Elec., 26,* 30, 116, 1921.
20. Angot, A., Series of articles on aurora, *Lumiere Elec., 8,* 1882.
21. *Angot, A., *The Aurora Borealis,* Kegan, Paul, Trench, Trubner, and Company, London, 1896.
22. Angström, A. J., Spectrum des Nordlichts, *Ann. Phys., 137,* 161, 1869.
23. Angström, A. J., Of the spectra of the aurora borealis, *J. Phys. Theor. Appl., 3,* 210, 1874.
24. Anonymous, *Annals of Cloon-Mac-Noise,* ~668 A.D. (First mention of aurora in Irish writings.)
25. Anonymous, *American Magazine and Monthly Chronicle for the British Colonies,* vol. 1, p. 25, Society of Gentlemen, Philadelphia, 1757.
26. Anonymous, *Under the Auroras — A Wonderful Story of the Interior World,* Exelsior, New York, 1888.

27. *Arago, D. F. J., *Meteorological Essays* (English translation), p. 389, Longman, Brown, Green, and Longmans, London, 1855.

28. Arctowski, H., *Aurores Australes, Results of the Voyage of S.Y. Belgica*, Buschmann, Brussels, 1902.

29. Aristotle, *Meteorologica*, Clarendon Press, Oxford, 1910–1930. (English translation of *Works* edited by W. D. Ross.)

30. Arnauld, H., *Aurore Boréale*, Gounouilhou, Bordeaux, 1913.

31. Arrhenius, S., La cause de l'aurore boreale, in *Revue General des Sciences*, p. 65, Armand Colin, Paris, 1902.

32. Arrhenius, S., On the electric charge of the sun, *Terr. Magn. Atmos. Elec., 10*, 1, 1905.

33. Arrhenius, S., *Die Nordlichter in Island und Grönland*, Almgvist and Wiksells, Uppsala, 1906.

34. Cavendish, R. (Ed.), Aurora, in *Man, Myth and Magic*, vol. 6, p. 175, McDonald Raintree Inc., Milwaukee, 1970.

35. Aytoun, W. E., *Lays of the Scottish Cavaliers*, W. Blackwood, London, 1863.

36. Babcock, H. D., A study of the green auroral line by the interference method, *Astrophys. J., 57*, 201, 1923.

37. *Backhouse, T. W., The spectrum of the aurora, *Nature, 28*, 209, 1883.

38. Baeblich, D., *Das Nordlicht*, Giegfried Gronbach, Berlin, 1871.

39. Banks, J., *The Endeavour Journal of Joseph Banks*, edited by J. G. Beaglehole, p. 149, Angus and Robertson, Sydney, 1962.

40. Baschin, O., Die ersten nordlichtphotographien, Aufgenommen in Bossekop (Lappland), *Meteorol. Z., 17*, 278, 1900.

41. Bates, D. R. (Ed.), *The Planet Earth*, Pergamon Press, London, 1957.

42. Bates, D. R., Auroral audibility, *Nature, 244*, 217, 1973.

43. Baumhauer, M. E. H., Sur l'origins des aurores polaires, *Comptes Rendus, 74*, 678, 1872.

44. Beals, C. S., The audibility of the aurora and its appearance at low atmospheric levels, *J. Roy. Astron. Soc. Can., 27*, 184, 1933.

45. Belknap, J., Extract of a letter from Reverend Jeremy Belknap containing observations on the aurora borealis, *Trans. Amer. Phil. Soc., 2*, 196, 1786.

46. Bell, J. M., The fireside stories of the Chippwyans, *J. Amer. Folklore, 16*, 73, 1903.

47. Bellinghausen, T., *The Voyage of Captain Bellinghausen to the Antarctic Seas, 1891-1921*, translated by F. Debenham, pp. 133, 147, Hakluyt Society, London, 1945.

48. Belon, A. E., J. E. Maggs, T. N. Davis, K. B. Mather, N. W. Glass, and G. F. Hughes, Conjugacy of visual auroras during magnetically quiet periods, *J. Geophys. Res., 74*, 1, 1969.

49. Benediktsson, E., Northern lights, in *Icelandic Poems and Stories*, translated by J. Honson, edited by R. Beck, Princeton University Press, Princeton, N. J., 1963.

50. Bequerel, H., in *Aurora, Their Characters and Spectra*, Spon, London, 1879.

51. Bernacchi, L. C., et al., *National Antarctic Expedition 1901-1904 — Physical Observations with Discussions by Various Authors*, Harrison and Sons, London, 1908.

52. Berry, R., Look north for aurora, *Astronomy, 4*, 43, 1976.

53. *Bertholon, A., Aurore Boreale, in *Encyclopedia Methodique 146, Dict. de Physique 1*, Academic Science, Paris, 1793. (Illustrations appeared in later edition (Agasse, Paris) on pages 139–148, 1816.)

54. Bierman, L., Kometenschweife und solare Korpuskularstrahlung, *Z. Astrophys., 29*, 274, 1951.

55. Biot, J. B., Consideration of the nature and origins of the aurora, (in French), *An. Phys., 67*, 1, 173, 1821. (With comments by Gilbert.)

56. *Birkeland, K., *The Norwegian Aurora Polaris Expedition, 1902-1903*, H. Aschehoug Co., Christiania, 1908.

57. Block, L. P., Scaling considerations for magnetospheric model experiments, *Planet. Space Sci., 15*, 1479, 1967.

58. Boas, F., The Eskimo of Baffin Land and Hudson's Bay, *Bull. Amer. Mus. Natur. Hist., 15*, 146, 1907.

59. Bóbrik, B. A., *Polarlicht und Spectral Beobachtungen*, vol. 2, part 1, Karl Gerold, Wien, 1886.

60. *Böller, W., Das südlicht, *Beitr. Geophys. 3*, 56–130, 550–609, 1898.

61. Botley, C. M., *The Air and Its Mysteries*, Bell and Sons, London, 1938.

62. *Botley, C. M., *Polar Lights*, Tunbridge Wells, England, printed by Courier Co., 1947. (Privately published.)

63. Botley, C. M., Halley and the aurora, *J. Brit. Astron. Ass., 66*, 31, 1955.

64. Botley, C. M., Some great tropical aurorae, *J. Brit. Astron. Ass., 67*, 188, 1957.

65. Botley, C. M., Aurora in SW Asia 1097–? 1300, *J. Brit. Astron. Ass., 74*, 293, 1963.

66. Botley, C. M., Some neglected aspects of the aurora, *Weather, 18*, 217, 1963.

67. Botley, C. M., Northern noises, *Weather, 19*, 270, 1964.

68. Botley, C. M., The aurora and the weather, *Weather, 20*, 117, 1965.

69. Botley, C. M., Unusual auroral periods, *J. Brit. Astron. Ass., 77*, 328, 1967.

70. Boué, A., Chronologischer Katalog der Nordlichter bis zum Jahre 1856, *Sitzungsber. Akad. Wiss. Wien, 22*, 3, 1857.

71. *Bradford, D., *The Wonders of the Heavens*, American Stationers Co., Boston, 1837.

72. Bradley, F., Extracts from an auroral register kept at New Haven, Connecticut, *Trans. Conn. Acad. Arts Sci., 1*, 139, 1866.

73. Bradley, L., Physics of the globe, 1, Aurora Borealis, *Proc. Amer. Ass. Advan. Sci., 19*, 82, 1860.

74. Bramhal, E. H., Goats, brands, and theories, *M.I.T. Tech. Rev., 46*, 2, 1943.

75. Bravais, A., *Voyages en Scandinaie, en Laponie, au Spitzberg et au Feröe Pendant les Annees 1838, 1839, 1840*, Bertrand, Paris, 1846.

76. Brecher, R., and E. Brecher, The secret of the aurora, *Readers Dig., 45*, 1954.

77. *Brewster, L., Aurora borealis, in *Edinburgh Encyclopedia*, vol. 3, p. 107, J. and E. Parker, Philadelphia, 1832.

78. *Briggs, J. M., Jr., Aurora and enlightenment, Eighteenth century explanations of the aurora borealis, *Isis, 58*, 491, 1967.

79. Brinton, D. G., *The Myths of the New World*, p. 268, Holt, New York, 1876.

80. Bronowski, J., *Science and Human Values*, Harper, New York, 1959.

81. Browning, R., *Globe Edition of Browning*, MacMillan, New York, 1921.

82. Bruun, D., *Kampen om Nordpolen*, p. 136, Ernst Bojesen, Copenhagen, 1902.

83. *Burch, J. L., The magnetosphere, in *Upper Atmosphere and Magnetosphere*, Geophysics Research Board of the National Academy of Sciences, Washington, D. C., 1978.

84. Burns, R., *The Complete Poetical Works of Robert Burns*, Houghton-Mifflin, Boston, 1897.

85. Bush, D., *Science and English Poetry: A Historical Sketch, 1590–1950*, p. 143, Oxford University Press, New York, 1950.

86. Byrd, R. E., *Alone*, Putnam and Sons, New York, 1938.

87. Byron, G., *The Complete Poetical and Dramatic Works of Lord Byron*, John Highland, Philadelphia, 1886.

88. Camden, W., *The True and Royal History of Elizabeth, Queen of England*, Purshue, Lownes and Flesher, London, 1625.

89. Canton, J., Electrical experiments with an attempt to account for their several phenomena, *Phil. Trans. Roy. Soc., 48*, 350, 1753.

90. *Capron, J. R., *Aurora, Their Characters and Spectra*, Spon, London, 1879.

91. Carlheim-Gyllensköld, V., *Exploration Internationale des Regions Polaires, 1882 to 1883*, vol. 2, part 1, Norstedt, Stockholm, 1886.

92. Carrington, R., Descriptions of a singular appearance seen in the sun on 1 September, 1859, *Mon. Notic. Roy. Astron. Soc., 20*, 13, 1860.

93. Cassini, J. D., Decouverte de la lumiere celeste qui paroist dans le zodiaque, *Mem. Acad. Roy. Sci., 8*, 179, 1666–1699.

94. Castro Medice, R., *Prognostication of the Great Signs that Appeared in the Sky*, printed with approval, Cordoba, 1605.

95. Cavendish, H., On the height of the luminous arch which was seen on February 23, 1784, *Phil. Trans. Roy. Soc., 80*, 101, 1790.

96. Celsius, A., Bemerkungen über der Magnetnadel Stündliche Veränderungen in ihrer Abweichung, *Svenska Ventensk. Handl., 8*, 296, 1747.

97. *Chamberlain, J. W., Theories of the aurora, in *Advances in Geophysics,* edited by H. E. Landsberg and J. Van Mieghem, p. 109, Academic Press, New York, 1958.

 *Chamberlain, J. W. *Physics of the Aurora and Airglow,* Academic Press, New York, 1961.

98. Chapman, S., Outline of a theory of magnetic storms, *Proc. Roy. Soc. London, Ser. A, 95,* 61, 1918.

99. Chapman, S., On solar ultra-violet radiation as the cause of aurorae and magnetic storms, *Mon. Notic. Roy. Astron. Soc., Geophys. Suppl., 2(6),* 296, 1930.

100. Chapman, S., A new theory of magnetic storms, *Nature, 126,* 129, 1930.

101. Chapman, S., The audibility and lowermost altitude of the aurora polaris, *Nature, 127,* 341, 1931.

102. Chapman, S., Polar and tropical aurorae, and the isoauroral diagram, *Proc. Indian Acad. Sci., 37,* 175, 1953.

103. *Chapman, S., History of aurora and airglow, in *Aurora and Airglow,* edited by B. M. McCormac, Reinhold, New York, 1967.

104. Chapman, S., Perspective, in *Physics of Geomagnetic Phenomena,* edited by S. Matsushita and W. H. Campbell, Academic Press, New York, 1967.

105. *Chapman, S., and J. Bartels, *Geomagnetism,* vol. 1, chap. 14, Clarendon Press, Oxford, 1940.

106. Chapman, S., and V. C. A. Ferraro, A new theory of magnetic storms, *Terr. Magn. Atmos. Elec., 36,* 77, 171, 1931; *37,* 147, 421, 1932.

107. Chase, P. E., On the relation of auroras to the gravitating currents, *Proc. Amer. Phil. Soc., 12,* 121, 1871.

 Chase, P.E., General relation of auroras to rainfall, *Proc. Amer. Phil. Soc., 12,* 400, 1872.

108. Chase, P. E., Influence of meteoric showers on auroras, *Proc. Amer. Phil. Soc., 12,* 401, 1872.

109. Chaucer, G., *The Canterbury Tales: The Knight's Tale,* Paul, Trench, Trübner, and Company, London, 1902.

110. Chree, C., Aurora and allied phenomena, *Quart. J. Roy. Meteorol. Soc., 49,* 67, 1923.

111. Clark, E. E., *Indian Legends of Canada,* p. 101, McClelland and Stewart, Toronto, 1960.

112. Claudianus, C., *Rape of Proserpine,* Book 1, lines 232–236, ~400 A.D.

113. Coleridge, S. T., *The Poetical Works of Samuel Taylor Coleridge,* MacMillan, London, 1909.

114. Collins, W., *The Poetical Works of William Collins,* edited by A. Dyce, Pickering Press, London, 1827.

115. Cook, J., *The Journal of Captain James Cook,* vol. 2, p. 95, Cambridge University Press, Cambridge, 1961.

116. Cooke, W. E., letter to *Nature, 81,* 525, 1909.

117. Corliss, W. R., *Handbook of Unusual Natural Phenomena,* Sourcebook Project, Glen Arm, Md., 1977.

118. Correspondent, Account of an aurora borealis, *Trans. Amer. Phil. Soc., 1,* 338, 1771.

119. *Cotté, P., *Sur l'Aurore Boréale, Memoires sur la Météorologie,* Royal Press, Paris, 1786.

120. Cox, A., Geomagnetic reversals, *Science, 163,* 237, 1969.

121. Cox, J., Comets' tails, the corona, and the aurora borealis, *Smithson. Inst. Annu. Rep., 1901–1902,* 179, 1903.

122. Craig, M. *Réfutation de la Loi de Gravitation de Newton,* Librairie Atar, Geneva, 1912.

123. Cranch, C. P., To the aurora borealis, *Dial, 1(1),* 10, 1840.

124. Crantz, D., *The History of Greenland,* pp. 202, 206, 233, Longman, Hurst, Rees, Orme, and Browne, London, 1820.

125. Cresswell, G., Fire in the sky, *Alas. Sportsman, 27,* Jan. 1968.

126. Cullington, A. L., A man-made or artificial aurora, *Nature, 182,* 1365, 1958.

127. Dall'Olmo, U., An additional list of auroras from European sources from 450 to 1466 A. D., *J. Geophys. Res., 84,* 1525, 1979.

128. Dalton, J., Essay on effects of the moon in producing the aurora borealis, in *Mathematical and Philosophical Repository,* edited by Davidson, 1789.
   *Dalton, J., *Meteorological Observations and Essays,* Richardson, Phillips, and Pennington, London, 1793.
129. Daniken, E., *Chariot of the Gods,* Putnam, New York, 1974.
130. Dauvillier, A., *Physique Cosmique,* vol. 4, *Le magnétisme des Corps Célestes, Le aurores polaires,* Hermann, Paris, 1954.
131. Davis, T. N., Probing the mysteries of the aurora, in *Britannica Yearbook of Science and the Future,* p. 144, 1974.
132. Davis, T. N., and E. M. Wescott, Active experiments in space, Annual Report, Geophys. Inst., Univ. of Alas., Fairbanks, 1974–1975.
133. Day, P., Meteorological Journal and lecture on Aurora, 1811; preserved at Yale University Library, see also E. Loomis, *Trans. Conn. Acad. Arts Sci., 1,* 169, 1866.
134. De la Rive, M. A., *A Treatise on Electricity in Theory and Practice,* p. 283, Longman, Brown, Green, Longmans, and Roberts, London, 283, 1858.
   De la Rive, M. A., New research on the aurora borealis and australis, *Arch. Sci. Phys. Natur., 14,* 122, 1862.
135. De la Rive, M. A., *Catalogue of the Special Loan Collection of Scientific Apparatus at the South Kensington Museum, 1876,* 3rd ed., p. 385, South Kensington Museum, London, 1877.
136. Derham, W., Observations on the Lumen Boreale, *Phil. Trans. Roy. Soc., 34,* 245, 1728.
137. Derr, J. S., Earthquake lights: A review of observations and present theories, *Bull. Seismol. Soc. Amer., 63,* 2177, 1973.
138. Desaguliers, M., Observations of electricity, *Phil. Trans. Roy. Soc., 42,* 14, 1742.
   Desaguliers, M., Conjectures concerning electricity, and the rise of vapours, *Phil. Trans. Roy. Soc., 42,* 140, 1742.
139. Descartes, R., *Vie et Oeuvres de Descartes,* edited by C. Adam, p. 197, Vrin, Paris, 1956.
140. *Devik, O., Kr. Birkeland as I knew him (memorial lecture), in *The Birkeland Symposium on Aurora and Magnetic Storms,* p. 13, edited by A. Egeland and J. Holtet, Centre National de la Recherche Scientifique, Paris, 1968.
141. Dickinson, E., *The Poems of Emily Dickinson,* edited by T. H. Johnson, Belknap and Cambridge, 1955.
142. Dionysius of Halicarnassus, *Roman Antiquities,* vol. 5, 46:2, ~20 B.C.
143. D.J.A.M.R.D.C., *Dissertations sur le Feu Boreal,* Joseph Bullot, Paris, 1733. (Initials were "pen name" for J. A. Mareusson.)
144. Donati, G.-B., Phenomena manifested in telegraphic lines during the great aurora borealis of February 4, 1872; and the origin of northern lights, Ann. Rep., Smithson. Inst. (1872), p. 268, Washington, D. C., 1873.
   Donati, G. B., I fenomeni luminosi della grande aurorae polare, *Mem. Osserv. Arcetri, 1*(1), 1873.
145. Drake, W. R., *Gods and Spacemen in Greece and Rome,* Signet Books, London 1976.
146. Eather, R. H., The Antarctic aurorae, *Austral. Photogr., 12,* 34, 1964.
147. Eather, R. H., Advances in magnetospheric physics: Aurora, *Rev. Geophys. Space Phys., 13,* 925, 1975.
148. Eddy, J. A., The Maunder minimum, *Science, 192,* 1189, 1976.
149. Editorial comment, On the noises that sometimes accompany the aurora borealis, *Edinburgh New Phil. J., 3,* 156, 1826.
150. *Edinburgh Rev. Critical J., The aurora borealis, *164,* 416, 1886.
151. Edlund, E., Recherches sur l'induction unipolaire, l'electricité atmospherique et l'aurore boréale, *Acta Acad. Sci. Suede, 16,* 1878.
152. Egede, H., *A Description of Greenland* (English translation from Danish), C. Hitch, London, 1745.
153. Egeland, A., and A. Omholt, Carl Störmer's height measurements of aurora, *Geophys. Publ., 26*(6), 1, 1966.
154. *Egeland, A., Ø. Holter, and A. Omholt (Eds.), Historic preamble, in *Cosmic Geophysics,* Universitetsforlaget, Oslo, 1973.

155. Eggers, C. U. D., *History of Iceland*, p. 245, August Friderich Stein, Copenhagen, 1786.

156. Eigen, M., *The Physicist's Concept of Nature*, edited by J. Mehra, Reidel, Boston, 1973.

157. Einstein, A., *The World as I See It*, translated by A. Harris, Watts and Company, London, 1940.

158. Eliot, T. S., The Lovesong of J. Alfred Prufrock, in *Poems by T. S. Eliot*, A. A. Knopf, New York, 1927.

159. Elliot, H., and J. J. Quenby, The Samoan artificial aurora, *Nature, 183*, 810, 1959.

160. Ellison, M. A., *The Sun and Its Influence*, Routledge and Kegan Paul, Ltd., London, 1943.

161. Elvey, C. T., and F. E. Roach, Aurora and airglow, *Sci. Amer., 193*, 140, 1955.

162. Elvey, C. T., and C. Rust, Can you hear the northern lights?, *Alas. Sportsman*, 18, June 1962.

163. Emme, E. M., *Aeronautics and Astronautics, 1915–1960*, U.S. Government Printing Office, Washington, D. C., 1961.

164. Ethier, V. G., The aurora, *Beaver, 9*, 4, 1968.

165. Euler, L., Recherches physiques sur la cause des queues des comètes, de la lumière boréale, et de la lumière zodiacale, *Hist. Acad. Roy. Sci. Belles Lett. Berlin, 2*, 117, 1746.

166. Eve, A. S., Northern lights, *Smithson. Inst. Annu. Rep. 1936*, 145, 1937.

167. Eyewitness, Account of aurora of Dec. 17, 1719, *Collect. Mass. Hist. Soc., 2*, 17, 1793.

168. Farquharson, J., Aurora borealis, *Edinburgh New Phil. J., 6*, 392, 1829.

    Farquharson, J., Report of a remarkable appearance of the Aurora Borealis below the clouds, *Edinburgh New Phil. J., 38*, 135, 1844.

169. Feerhow, F., *The Influence of Earth Magnetism on Man, With a Theory of the Northern Lights* (in German), Altman, Leipzig, 1912.

170. Feldstein, Y. I., Some problems concerning the morphology of auroras and magnetic disturbances at high latitudes, *Geomagn. Aeron., 3*, 183, 1963.

171. *Ferraro, V. C. A., Aurora and magnetic storms, in *The Planet Earth*, edited by D. R. Bates, Pergamon Press, New York, 1957.

172. Foran, J. K., *Poems and Canadian Lyrics*, Sadlier and Sons, Montreal, 1899.

173. Force, P., Record of meteorological phenomena observed in the higher northern latitude, *Smithson. Contrib. Knowl.*, 1856.

174. Forssman, L.-A., *The Relations of the Aurora Borealis to Magnetic Perturbations and Meteorological Phenomena* (in French), E. Berling, Upsul, 1872.

175. Frank, L., Interview in film "Earthspace—the Magnetosphere," produced by R. H. Eather and B. Kaufman, Boston, 1976. (See list of films about the aurora.)

176. Franklin, B., *Political, Miscellaneous and Philosophical Pieces*, edited by Vaughan, p. 504, Johnson, London, 1779.

177. Franklin, J., *Narrative of a Journey to the Shores of the Polar Sea in the Years 1819, 1920, 1921, and 1922*, M. G. Hurtig Ltd., Edmonton, 1975.

178. Fritsche, M., *Meteororum . . . Item Catalogus Prodigiorum*, Montani, Noribergae, 1563.

179. *Fritz, H., *Das Polarlicht*, Brockhaus, Leipzig, 1881.

180. *Frobesius, J. N., *Modern and Ancient Spectacular Appearances of the Aurora Borealis with a Philosophical Analysis of the Remarkable Phenomenon*, Weygandum, Helmstadt, 1739. (Also contains auroral catalog, 502 B.C.–1739 A.D.)

181. Fulke, W., *A Goodly Gallerye With a Most Pleasant Prospect, Into the Garden of Naturall Contemplation*, Griffith, London, 1563.

182. Gabor, D., *Inventing the Future*, Secker and Warburg, London, 1963.

183. Galileo, G., *Le Opere di Galileo Galilei*, 2nd ed., Edizione Nazionale, Florence, 1929–1939.

184. Gannon, R., How science is solving the mystery of the northern lights, *Pop. Sci., 203*, 82, 1973.

185. Garber, C. M., *Stories and Legends of the Berring Straight Eskimos*, Christopher Publishing House, Boston, 1940.

186. Garkavi, A. J., *Tales by Moslem Writers About the Slavs and Russians*, p. 88, Saint Petersburg, 1870.

187. *Gartlein, C. W., Unlocking the secrets of the northern lights, *Nat. Geogr., 92,* 673, 1947.

188. Gartlein, C. W., Auroral spectra showing broad hydrogen lines, *Eos Trans. AGU, 31,* 18, 1950.

189. Gartlein, C. W., U.S. visual observations, *IGY Newslett., 26,* 1959.
    Gartlein, C. W., Visual observations of the aurora, in *Geophysics and the IGY,* edited by H. Odishaw and S. Ruttenberg, American Geophysical Union, Washington, D. C., 1958.

190. Gassendi, P., The aurora borealis which is seen to glow and move at night, in *Opera Omnia,* vol. 2, p. 117, Montmort, Lyons, 1658.

191. Geddes, A. E. M., *Meteorology,* Blackie and Son, London, 1921.

192. Gehler, J. S. T., *Physikalisches Wörterbuch,* E. B. Schwickert, Leipzig, 1837.

193. Geinzius, article, in *Primecanija na Vedomosti v St. Petersburg,* 1740.

194. *Gemma, C., *De Naturae Divinis Characterismis,* Antwerpiae, 1575.

195. *Gentlemen's Mag., 20,* 418, 1750; *21,* 39, 1751.

196. Gmelin, J. G., *Voyage en Siberia,* vol. 2, p. 31, M. de Keralio, Paris, 1767.

197. Godin, M., Sur le météore qui aparu le 19 Octobre de cette année, *Mem. Acad. Roy. Sci.,* 287, 1726.

198. Goethe, W., *Lyrische Epische Dischtungen, 1,* 226, 1920.

199. Gold, T., Motions in the magnetosphere of the earth, *J. Geophys. Res., 64,* 1219, 1959.

200. Goldstein, E., Über der electrizität in verdünnten Gasen., *Ann. Phys. Chem., 12,* 266, 1881.

201. Gottlieb, M. B., Plasmas, *Phys. Today, 21,* 46, 1968.

202. Gray, C. H. (Ed.), *Mythology of All Races,* vol. 10, p. 35, Cooper Square, New York, 1964.

203. Greely, A. W., *Three Years of Arctic Service,* Charles Scribner Sons, New York, 1886.

204. *Green, S. A., Remarks on the early appearance of the northern lights in New England, in *Proceedings of the Massachusetts Historical Society,* J. Wilson, Cambridge, 1885.

205. Greenwood, I., Of an aurora borealis seen in New England, *Phil. Trans. Roy Soc. London, 37,* 55, 1731.

206. Gregory of Tours, in *Patrologie Cursus Completus,* vol. 71: *Gregorii Turnonousis Episcopi Opera Omnia,* edited by J. P. Migne, Garnier, Paris, 1849.

207. Gromnica, E., An unbelievable story about the northern lights, *North, 18,* 44, 1971.

208. *Günther, S., Polarlicher, *Handl. Geophys., 1,* 588, 1897.

209. *Haavio, M., Folklore of the northern lights, *Minutes Finn. Acad. Sci. Lett.,* 149, Dec. 15, 1943.

210. Halley, E., An account of the late surprising appearance of the lights seen in the air, *Phil. Trans. Roy. Soc., 29,* 406, 1716.

211. Halley, E., An account of the phenomena of a very extraordinary aurora borealis, *Phil. Trans. Roy. Soc., 30,* 1099, 1719.

212. Hamilton, J. C., The Algonquin Manabozho and Hiawatha, *J. Amer. Folklore, 16,* 231, 1903.

213. Hansteen, C., On the aurora borealis and polar fogs, *Edinburgh Phil. J., 12,* 83, 235, 1825.
    Hansteen, C., On the polar lights, or aurora borealis and australis, *Phil. Mag., 2,* 334, 1827.

214. *Harang, L., *The Aurorae,* John Wiley and Sons, New York, 1951.

215. Haüy, M. R. J., *Natural Philosophy,* vol. 2, translated by O. Gregory, G. Kearsley, London, 1807.

216. Hawkes, E. W., The Labrador Eskimo, *Mem. 91,* p. 137, 153, Geol. Surv. of Can., Ottawa, 1916.

217. H. C. B., An artful and amusing attempt at alphabetical alliteration addressing aurora, *Notes and Queries, Ser. 2, 8,* 412, 1859.

218. Hearne, S., *Journey From Prince of Wales Fort in Hudson's Bay to the Northern Ocean,* MacMillan, Toronto, 1958.

219. Heath, T. L., *Greek Astronomy*, J. M. Dent and Sons, London, 1932.
220. Heberer, L., Nordlicht!, *Reclams Universum, 14*, 1, 1927.
221. Heikkila, W. J., Aurora, *Eos Trans. AGU, 54*, 764, 1973.
222. Heitman, J., *Physical Considerations About the Sun's Heat, Cold Air, and the Northern Lights*, Y. Y. Hopffner, Oslo, 1741.
223. Hellman, G., *Repertorium der Deutschen Meteorologie*, Engelmann, Leipzig, 1883.
224. *Hellman, G., Die älteste gedruckte Nordlichtbeschreibung, Beiträge zur Geschichte der Meteorologie, *Veröff Königlich* Preuss. *Meteorol. Inst., 273*(3), 107, 1914. (The oldest printed auroral description.)
   Hellman, G., *Die Meteorology in den Deutschen Flugschriften und Flugblattern des XVI Jahrhumderts*, Akademy der Wissenschaffen, Berlin, 1921.
225. Henriksen, R. N., and D. R. Rayburn, Hot pulsar magnetospheres, *Mon. Notic. Roy. Astron. Soc., 166*, 409, 1974.
226. Herrick, E. C., A register of the aurora borealis at New Haven, Connecticut, *Trans. Conn. Acad. Arts Sci., 1*, 7, 1866.
227. Hesiod, *Theogony*, 183–184, 689–693, 824–828, eighth century B.C.
228. *Hess, W. N., *The Radiation Belt and Magnetosphere*, Blaisdell Publishing Co., Waltham, Mass., 1968.
229. Hess, W. N., Generation of an artificial aurora, *Science, 164*, 1512, 1969.
230. Heyl, P. R., *Physics*, edited by S. Rapport and H. Wright, p. 90, New York University Press, New York, 1964.
231. Hindle, B., *The Pursuit of Science in Revolutionary America, 1735–1789*, University of North Carolina Press, Chapel Hill, 1956.
232. Hindley, K., The northern lights unveiled, *New Sci., 40*, 74, 1937.
233. Hiorter, O. P., Von der magnetnadel verschiedenen bewegungen, *Sv. Vetensk. Handl., 8*, 27, 1747.
234. Hodgson, R., On a curious appearance seen on the sun, *Mon. Notic. Roy. Astron. Soc., 20*, 16, 1860.
235. Holden, A. P., The sunspot cycle, *Observatory, 2*, 389, 1879.
236. Hole, C. (Ed.), *Encyclopedia of Superstition*, p. 26, Hutchinson, London, 1961.
237. Holm, G., Ethnological sketch of the Angmagssalik Eskimos, *Medd. Grøenland, 39*, 82, 1914.
238. Holyoke, E. A., *Memoirs of Dr. Holyoke*, p. 27, Essex South District Medical Society, Boston, 1829. (Also in Felt's *History of Salem*, vol. 2, p. 137, 1829.)
239. *Holzworth, R. H., Folklore and the aurora, *Eos Trans. AGU, 56*, 686, 1975.
240. Hopkins, G. M., *Poems and Prose of Gerard Manley Hopkins*, p. 123, Penguin Books, New York, 1953.
241. Hulburt, E. O., The origin of the aurora borealis, *Terr. Magn. Atmos. Elec., 33*, 11, 1928.
242. Humboldt, A., *Sketch of a Physical Description of the Universe*, vol. 3, p. 41, Longman, Brown, Green, and Longmans, London, 1847.
243. Hunsucker, R. D., The northern lights, *Alas. Sportsman*, 8, March 1963.
244. Isaev, S. I., and N. V. Puskov, *Aurora*, Akademija Nauk SSSR (Popular Science Series), Moscow, 1958.
245. *Jacka, F. (Ed.), *International Auroral Atlas*, University Press, Edinburgh, 1963.
246. Jacob, V., The Northern Lights, in *Holyreed—A Garland of Modern Scots Poems*, edited by W. H. Hamilton, J. M. Dent and Sons, London, 1929.
247. Jeremiah, J., Early mention of the aurora borealis, *Nature, 3*, 174, 1870.
248. *J. F. R., Historical and physical accounts of the northern lights, *Acta Soc. Hafniensis*, p. 317, 1745.
249. Johnson, J. H., *Concerning the Aurora Borealis*, Gazette Press, Berkeley, 1930.
250. Jones, W., (untitled article), *J. Amer. Folklore, 24*, 214, 1911–1912.
251. Josephus, F., *The Genuine Works of Flavius Josephus*, vol. 6, chap. 5, translated by W. Whiston, Allason, London, 1818.
252. *Kaemtz, L. F., *A Complete Course of Meteorology*, translated by C. V. Walker, p. 446, Epolyte Bailliére, London, 1845.
253. Kaemtz, L. F., *Treatise on Meteorology* (in German), Gebauersche Buchhandlung, Halle, 1836.
254. Kawai, N., and K. Hirooka, Wobbling motion of the geomagnetic dipole field in historic time during these 2000 years, *J. Geomagn. Geoelec, 19*, 217, 1967.

255. Keats, J., *Complete Poems and Selected Letters,* edited by C. Thorpe, Doubleday, New York, 1935.

256. Keimatsu, M., *Documentary Catalog of Northern Lights Observed in China, Korea and Japan from 7 B.C. to 10 A.D.,* College of Liberal Arts, Kanazawa Univ., Kanazawa, Japan, 1965. (A chronology of aurorae and sunspots observed in China, Korea and Japan, part 1, *Ann. Sci. Kanazawa Univ., 7,* 1970; part 2, A.D. 1–3 century, *8,* 1971; part 3, 4–5 century, *9,* 1972; part 4, 6–8 century, *10,* 1973; part 5, A.D. 801–1000, *11,* 1974; part 6, A.D. 1001–1130, *12,* 1975; part 7, A brief summary of records from B.C. 687 to A.D. 1600, *13,* 1976.)

257. Keimatsu, M., N. Fukushima, and T. Nagata, Archaeo-aurora secular variation in historic time, *J. Geomagn. Geoelec., 20,* 45, 1968.

258. Kennan, G., *Tent Life in Siberia,* G. P. Putney, New York, 1870.

259. Kerillis, C.-A., *L'Aurore Boréale,* Gustave Flicker, Paris, 1911.

260. Kilgour, F. G., Professor John Winthrop's notes on sun-spot observations (1739), *Isis, 30,* 473, 1939.

261. *Kimball, D. S., A study of the aurora of 1859, *Rep. 6,* Geophys. Inst., Univ. of Alas., Fairbanks, 1960.

262. King, W. F., Audibility of the aurora, *J. Roy. Astron. Soc. Can., 1,* 193, 1907.

263. Kirwan, R., On aurora borealis, *Devonshire Ass. Advan. Sci. Lit. Art Rep. Trans., 5,* 344, 1872.

264. Kleinbaum, A. R., Jean Jacques Dortous De Mairan; A study of an enlightenment scientist, Ph.D. thesis, Columbia Univ., New York, 1970.

265. Koch, K. R., *Resultate der Polarlicht—Beobachturgen Winter 1882–1883* Asher Publishing, Berlin, 1885.

266. Kochen, in *The Russian Old Times,* vol. 9, p. 125, 1878.

267. Kotzschmar, H., Aurora borealis, mazurka for piano, H. Kotzchman, New York, 1852.

268. Kraft, G. W., Observationum Meteorologum ab 1726 usque in finem anni 1736, *Comment. Acad. Sci. Imper. Petropolitanae, 9,* 316–363, 1739.

269. Krauss, F. B., *An Interpretation of the Omens, Portents and Prodigies Recorded by Livy, Tacitus and Suetonius,* University of Pennsylvania Press, Philadelphia, 1930.

270. Kroeber, A. L., The Eskimos of Smith Sound, *Bull. Amer. Mus. Natur. Hist., 12,* 319, 1899.

271. Kulmus, J. A., *Detailed Description of the Northern Lights Observed on March 1, 1721,* J. D. Stollen, Danzig, 1721.

272. Kurilov, G. N., private communication to R. H. Eather, Institute of Language, Literature and History, Yakutsk, Russia, 1976.

273. *Larson, L. M. (translator), *The King's Mirror,* Twayne Publishing, New York, 1917.

274. Layamon, *Arthurian Chronicles,* edited by Wace and Layamon, Dent and Sons, London, 1921.

275. Leach, M. (Ed.), *Standard Dictionary of Folklore, Mythology and Legend,* Funk and Wagnals, New York, 1949.

276. Lefroy, R. A., Report on the observations of the aurora borealis, made by the N. C. officers of the royal artillery, at the various guard-rooms in Canada, *Phil. Mag., 4,* 59, 1852.

277. LeGalley, D. P., and A. Rosen (Eds.), *Space Physics,* John Wiley, New York, 1967.

278. Lemström, S., *Catalogue of the Special Loan Collection of Scientific Apparatus at the South Kensington Museum, 1876,* 3rd ed., p. 385, South Kensington Museum, London, 1877.

279. *Lemström, S., *L'Aurore Boréale,* Gauthier-Villars, Paris, 1886.

280. Lemström, S., Luminous phenomena of the same kind as the auroral light, natural and artificial, *Ofyersigt Finska Vetensk. Soc. Forhandl., 41,* 32, 1899.

281. Lewis, R., An account of the same aurora borealis, *Phil. Trans. Roy. Soc. London, 37,* 69–70, 1731.

282. Libes, A., *Traité de Physique Noveau and Dictionnaire de Physique,* 4 volumes, Gigueta et Michaud, Paris, 1806.

283. Lindemann, F. A., Note on the theory of magnetic storms, *Phil. Mag., 38,* 669, 1919.

284. *Link, F., On the history of the aurora borealis, in *Vistas in Astronomy,* vol. 9, edited by A. Beer, Pergamon Press, New York, 1957.

285. Link, F., Observations et catalogue des aurores boréales apparues en occident de 626 A.D. a 1600 A.D. et 1601 a 1700, *Contrib. Inst. Astron. Acad. Tchec. Sci. Prague, 173,* 297, 1962.

286. Little, D. E., and G. M. Shrum, Correlation of auroral observations in the northern and southern hemispheres, *Trans. Roy. Soc. Can., Ser. 3, 44,* 51, 1950.

287. Lomonosov, M. V., *Rhetoric,* translated by W. N. Vickery, Academy Nauk, St. Petersburg, 1748.

288. Lomonosov, M. V., Oration on aerial phenomena, proceeding from the force of electricity, 1753, in *Russias Lomonosov,* translated by W. C. Huntington, Princeton University Press, Princeton, N. J., 1952.

289. Lomonosov, M. V., *Complete Collection of Papers of Lomonosov,* vols. 3, 4, USSR Academy of Sciences, Moscow, 1954.

290. Longfellow, H. W., *The Song of Hiawatha,* Tickner and Fields, Boston, 1885.
Longfellow, H. W., *Tales of a Wayside Inn,* Tickner and Fields, Boston, 1863.

291. Longfellow, H. W., Driftwood-Frithiof's Saga, in *The Writings of Henry Wadsworth Longfellow,* vol. 2, Houghton-Mifflin, Boston, 1886.

292. Loomis, E., On the geographical distribution of auroras in the northern hemisphere, *Amer. J. Sci. Arts, 30,* 89, 1860.

293. *Loomis, E., The aurora borealis, or polar light: Its phenomena and laws, *Smithson. Inst. Annu. Rep. 1864,* 208, 1865.

294. Loomis, E., Notices of auroras extracted from the meteorological journal of Reverend Ezra Stiles, *Trans. Conn. Acad. Arts Sci., 1,* 155, 1866–1871.

295. Lottin, V., A. Bravais, C. B. Lillcechöök, and P. A. Siljeström, *Voyages en Scandivaie en Laponie au Spitzberg et aux Feröe (1838, 1839 et 1840),* Bertrand, Paris, 1843.

296. Lovering, J., On the secular periodicity of the aurora borealis, *Mem. Amer. Acad. Arts Sci., 9,* 101, 1867.

297. Lovering, J., On the periodicity of the aurora borealis, *Mem. Amer. Acad. Arts Sci., 10,* 9, 1868–1871.

298. Lowe, E. J., *A Treatise on Atmospheric Phenomena,* Longman, Brown, Green, and Longmans, Nottingham, 1846.

299. Lucanas, M., *The Civil War,* Book 1, lines 522–534, ~60 A.D.

300. Lui, A. T. Y., C. D. Anger, D. Venkatesan, and W. Sawchuk, The topology of the auroral oval as seen by the Isis 2 scanning auroral photometer, *J. Geophys. Res., 80,* 1795, 1975.

301. *Lundmark, B., The audible light: Lapp folktales concerning aurora, *Vasterbotten, 1/2,* 86, 1976.

302. *Lycosthenes, C., *Prodigiorum ac Ostentorum Chronicon,* H. Petri, Basileae, 1557.

303. MacCulloch, C. J. A. (Ed.), *Mythology of All Races,* vol. 3, p. 319; vol. 4, pp. 79, 81–82, 287, 398, 488; vol. 10, pp. 35, 249, Marshall Jones, Boston, 1927.

304. Mackay, A. L., *The Harvest of a Quiet Eye,* Institute of Physics, London, 1977.

305. Mackenzie, D. A., *Scottish Folklore and Folklife,* p. 222, Blackie and Son, London, 1935.

306. *Mairan, J. J., *Traité Physique et Historique de l'Aurore Boréale,* 1st ed., L'Imprimerie Royale, Paris, 1733; 2nd revised edition, 1754.
Mairan, J. J., Conjectives sur l'origin de la fable de l'Olympie, in *Lettres au R. R. Parrenin,* p. 183, L'Imprimerie Royal, Paris, 1770.

307. Ma, Kuo-Han (Ed.), *Yu Han Sha Fang Chi I Shu* [The Jade-Box Mountain, Studio Collection of (reconstituted) Lost Books], 1853.

308. Manger, M., *Description of Aurora Seen in Bohemia,* Augsburg, 1510. (Legend attached to figure on p. 45.)

309. Maraldi, M. Observations d'une lumiere septentrionale, *Mem. Acad. Roy. Sci.,* 95–107, 1716; 22–31, 1717.

310. Maris, H. B., and E. O. Hulbert, A theory of aurora and magnetic storms, *Phys. Rev., 33,* 412, 1929.

311. Marsh, B. V., The aurora, viewed as an electric discharge, *Amer. J. Sci., 31,* 311, 1861.

312. *Mather, C., *A Voice From Heaven,* Samuel Kneeland, Boston, 1719.

313. Mathew of Westminster, *Annals of Philosophy,* vol. 9, p. 250, ~555 A.D.

314. Matsushita, S., Ancient aurora seen in Japan, *J. Geophys. Res., 61,* 297, 1956.

315. *Maunder, E. W., The prolonged sunspot minimum, 1645-1715, *J. Brit. Astron. Ass., 32,* 140, 1922.

316. Mawson, D., Auroral observations at the Cape Royd Station, Antarctica, *Trans. Roy. Soc. Sci. S. Aust., 40,* 151, 1916.
    Mawson, D., Records of the aurora polaris, *Scientific Report of the Australian Antarctic Expedition, 1911-1914,* Alfred Kent, Sydney, 1925.

317. Mayero, F. C., De Luce Boreali, *Comment. Acad. Sci. Imp. Petropolitanae, 1,* 121, 1729.

318. *McAdie, A., *The Aurora in Its Relations to Meteorology,* vol. 18, *Signal Service Notes,* War Department, Washington, D. C., 1885.

319. *McAdie, A., What is an aurora?, *Century Mag., 59,* 1897.

320. McGrath, W. C., *Fatima or World Suicide,* Scarboro Foreign Mission Society, Ontario, 1950.

321. McIlwain, C. E., Direct measurement of particles producing visible auroras, *J. Geophys. Res., 65,* 2727, 1960.

322. McKeehan, L. W., *Yale Science, The First Hundred Years, 1701-1801,* Henry Schuman, New York, 1947.

323. McLennan, J. C., and G. M. Shrum, On the origin of the auroral green line 5577 Å and other spectra associated with the aurora borealis, *Proc. Roy. Soc., Ser. A, 108,* 501, 1926.

324. Meek, J. H., *Science, History and Hudson Bay,* vol. 2, *Upper Atmospheric Research I. Aurora and Ionosphere,* edited by C. S. Beals, p. 729, Energy Mines Research, Ottawa, 1968.

325. Meinel, A. B., Doppler-shifted auroral hydrogen emission, *Astrophys. J., 113,* 50, 1951.

326. *Mendillo, M., and J. Keady, Watching the aurora from colonial America, *Eos Trans. AGU, 57,* 485, 1976.

327. Meredith, L. H., L. R. Davies, J. P. Heppner, and O. E. Berg, Rocket auroral investigations, in *Experimental Results of the U.S. Rocket Program for the IGY to 1 July, 1958,* vol. 1, *IGY Rocket Rep. Ser.,* edited by J. Hanessian and I. Guttmacker, p. 169, National Academy of Sciences, Washington, D. C., 1958.

328. Meyer, F. C., *De Luce Boreali,* vol. 1, p. 351, Academy of Science, Saint Petersburg, 1728.
    Meyer, F. C., *De Luce Boreali,* vol. 4, p. 121, Academy of Science, Saint Petersburg, 1728.

329. Milevskogo, O. N. (translator), *Journal of the Last Expedition of Stefan Batorija in Russia,* Pskovsk, Archedogich, 1882.

330. Milman, H. H., *The Poetical Works of the Reverend H. H. Milman,* J. Murray, London, 1839.

331. Milman, H. H., *History of the Jews,* vol. 1, Widdleton, New York, 1877.

332. Milton, J., *Paradise Lost,* vol. 8, P. Parker, R. Butler and M. Walker, London, 1667.

333. Minge, J. P., *Patrologiae Curses Compl. Ser. Lat.,* vol. 5, p. 283, Paris, 1883.

334. Monge, M., *Lecons de Physique,* p. 237, Pujoulz, Paris, 1805.

335. Morgan, J. H., and J. T. Barber, *An Account of the Aurora Borealis Seen Near Cambridge, October 24, 1847,* McMillan, Barclay and McMillan, Cambridge; G. Bell, London, 1848. (With 12 color engravings.)

336. Morphew, J., An essay concerning the late apparition in the heavens, *Roy. Soc. London,* 1716. (Submitted but apparently never published.)

337. Morrell, G. F., The finest aurora for many years: A phenomenon explained, *Illus. London News,* Jan. 1938.

338. Mottelay, P. F., *Bibliographical History of Electricity and Magnetism,* Charles Griffin and Company, London, 1922.

406. Schove, D. J., The sunspot cycle, 649 B.C.-2000 A.D., *J. Geophys. Res., 60,* 127, 1955.

407. Schove, D. J., Auroral numbers since 500 B.C., *J. Brit. Astron. Ass., 72,* 30, 1962.

408. Schove, D. J., and P. Y. Ho, Chinese aurorae: I, A.D. 1048-1070, *J. Brit. Astron. Ass., 69,* 295, 1959.

   Schove, D. J., and P. Y. Ho, Chinese records of sunspots and aurorae in the fourth century A.D., *J. Amer. Orient. Soc., 87,* 105, 1967.

409. Schröder, W., Über die Haüfigkeit der Polarlichter in Deutschland, *Gerlands Beitr. Geophys., 75,* 345, 1966.

   Schröder, W., Herman Fritz und sein Wirken für die Polarlichtforschung, in *Geschichte der Geophysik,* Springer, Berlin, 1974.

410. Schuster, A., The origin of magnetic storms, *Proc. Roy. Soc. London, Ser. A, 85,* 61, 1911.

411. Schwarz, H. T., *Elik, and Other Stories of the MacKenzie Eskimos,* McClelland and Stewart, Toronto, 1970.

412. Scott, R. F., *Scott's Last Expedition, Journal of Scott,* p. 284, arranged by L. Huxley, Smith, Elder, and Company, London, 1913.

413. Scott, W., *The Lay of the Last Minstrel,* William Smith, London, 1839.

414. Seneca, *Naturales Quaestiones,* Book 1, translated by T. H. Corcoran, Harvard University Press, Cambridge, Mass., 1972.

415. Serantoni, G. M., *Dialog Intorno All Cagione della Celebre Aurora Boreale December 16, 1737,* Lucca, Italy, 1740.

416. Service, R., *Complete Poems of Robert Service,* Dodd, Mead, and Company, New York, 1947.

417. Seydl, A., A list of 402 northern lights observed in Bohemia, Moravia and Slovakia from 1013 to 1951, *Geofys. Sb., 17,* 159, 1954.

418. Shackleton, E. H., *Aurora Australis,* manuscript 722, Scott Polar Research Institute, Cambridge, England.

419. Shackleton, E. H. (Ed.), *South Polar Times 1, 1902,* Smith, Elder, and Company, London, 1907.

420. Shackleton, E. H., *The Heart of the Antarctic,* vol. 1, William Heinemann, London, 1909.

421. Shafer, Yu. G., private communication to R. H. Eather, Institute of Cosmophysical Research and Aeronomy, Yakutsk, 1976.

422. Shakespeare, W., *A Midsummer Night's Dream,* Act 5, scene 1, line 14; *Julius Caesar,* Act. 2, scene 2, line 19.

423. Silberman, M. S.-J., Sur les aurores boréales, *Comptes Rendus, 68,* 1120, 1869.

   Silberman, M. S.-J., Sur les aurores boreales, en particulier, sur celles des 13, 14, et 15 Mai, 1869, *Comptes Rendus, 68,* 1164, 1869.

424. Silius Italicus, *Punica,* Book 1, lines 460-467; Book 8, lines 626-637, ~95 A.D.

425. Silverman, S., Franklin's theory of the aurora, *J. Franklin Inst., 290,* 177, 1970.

426. *Silverman, S. M., and T. F. Tuan, Auroral audibility, *Advan. Geophys., 16,* 155, 1972.

427. Singer, S. F., A new model of magnetic storms and aurorae, *Eos Trans. AGU, 38,* 175, 1957.

428. Singer, S. F., letter to *Sci. Amer., 209,* 8, 1963.

429. Sirr, H. C., *Ceylon and the Cingalese,* vol. 2, p. 117, W. Shoberl, London, 1850.

430. *Siscoe, G. L., Solar-terrestrial relations: Stone age to space age, *Technol. Rev., 79,* 26, 1976.

431. Siscoe, G. L., A historical footnote on the origin of aurora borealis, *Eos Trans. AGU, 59,* 994, 1978.

432. Smart, C., *A Song to David and Other Poems,* edited by R. Todd, Grey Walls Press, London, 1947.

433. Smith, D. M., *Arctic Expeditions from British and Foreign Shores,* Charles H. Calvert, Southhampton, 1877.

434. Smyth, P., Spectroscopic observations of the zodiacal light in April 1872 at the Royal Observatory Palermo, *Mon. Notic. Roy. Astron. Soc., 32,* 277, 1972.

435. Smyth, P., Astronomical observations made at the Royal University, Edinburgh, during the years 1870–1877, *Mem. Roy. Observ. Edinburgh, 14* (2), 29, 1878.

436. Solzhenitsyn, A. I., *One Day in the Life of Ivan Denisovich,* Dutton, New York, 1963.
    Solzhenitsyn, A. I., Matryana's House, in *Halfway to the Moon—New Writings From Russia,* Anchor-Doubleday, New York, 1965.

437. Southey, R., *Poems of Robert Southey,* edited by M. H. Fitzgerald, Oxford University Press, New York, 1909.

438. Sparrman, A., *A Voyage Around the World with Captain Cook in H. M. S. Resolution,* chap. 22, Golden Cockerel Press, London, 1944.

439. Steiger, W. R., and S. Matsushita, Photographs of the high-altitude nuclear explosion Teak, *J. Geophys. Res., 65,* 545, 1960.

440. Stevens, W., *Collected Works of Wallace Stevens,* Knopf Publishers, New York, 1965.

441. *Stewart, B., Aurora borealis, in *Encyclopedia Britannica,* vol. 16, 9th ed., p. 181, A. C. Black, Edinburgh, 1883.

442. Stiles, E., Notices of auroras extracted from the meteorological journal of Rev. Ezra Stiles, collected by Loomis, *Trans. Conn. Acad. Arts Sci., 1,* 155, 1866.

443. Stokes, F. W., The aurora borealis, *Century Mag., 65,* 486, 1903.

444. Störmer, C., Sur les trajectoires des corpuscles electrisés dans l'espace sous l'action du magnétisme terrestre avec l'application aux aurores boréales, *Arch. Sci. Phys. Natur., 24,* 5, 113, 221, 317, 1907.

445. Störmer, C., Progress in the photography of the aurora borealis, *Terr. Magn. Atmos. Elec., 37,* 475, 1932.

446. Störmer, C. (Ed.), *Photographic Atlas of Auroral Forms,* A. W. Broggers Boktrykkeri, Oslo, 1930.

447. *Störmer, C., *The Polar Aurora,* Clarendon Press, Oxford, 1955.

448. *Stothers, R., Ancient aurora, *Isis, 70,* 85, 1979.
    Stothers, R., Solar activity cycle during classical antiquity, *Astron. Astrophys., 77,* 121, 1979.

449. *Sturgeon, W., Description of several extraordinary displays of the aurora, with theoretical remarks, *Edinburgh New Phil. J., 47,* 147, 225, 1849.

450. *Sullivan, W., *Assault on the Unknown,* McGraw-Hill, New York, 1968.

451. Suxomlinova, M. I., *Collection of Works of Lomonosov, With Explanations and Notes,* vol. 4, p. 296, Saint Petersburg, 1898.

452. Sverdrup, H. U., Audibility of the aurora polaris, *Nature, 155,* 1941.

453. *Svjatskij, D. O., *Astronomical Phenomena in Russian Chronicles,* Academy of Science, Saint Petersburg, 1915.

454. *Svjatskij, D. O., *Aurora in Russian Literature and Science From the 10th to 18th Centuries,* Academy of Science of USSR, Leningrad, 1934.

455. Swan, J. G., Indians of Cape Flattery, *Smithson. Contrib. Knowl., 220,* 87, 1868.

456. Synge, J. M., *The Playboy of the Western World,* Maunsel and Company, Dublin, 1907.

457. Szent-Gyorgyi, A., Perspectives in biology and medicine, in *The Harvest of a Quiet Eye* by A. L. Mackay, 1971.

458. Tacitus, P. C., *Germania,* Oxford translation, Harper and Brothers, New York, 1898.

459. *Tai, N.-T., Ancient records of aurora in China and their contribution to science, *Kexue Tongbao, 20,* 457, 1975. (English summary in *New Scientist, 69,* 233, 1976. Original source, *Guankui Jiyao* (Summary of Sights Through a Peephole), edited by H. Ding, 1652.)

460. Tatiscev, V. N. (Tatishchev), *Russian History,* Moscow University Publications, Moscow, 1768.

461. Taylor, B., *Northern Travels,* G. P. Putman, New York, 1858.

462. Taylor, B. F., *Complete Poetical Works of Benjamin F. Taylor,* Griggs and Company, Chicago, 1886.

463. Tegner, E., *Frithjofssage,* canto 14, Norstedt, Stolkholm, 1831.

464. Tennyson, A., *The Poetical Works of Alfred Lord Tennyson,* Nims and Knight, New York, 1885.

465. Thomas, S., *Men of Space,* vol. 2, p. 218, Chilton Press, Philadelphia, 1961.

466. Thomson, J., *James Thomson Poetical Work,* edited by J. C. Robertson, Oxford University Press, New York, 1965.

467. Tindale, R., Private Communication to R. H. Holzworth, *Eos Trans. AGU, 56,* 688, 1975.

468. Todd, E., A Canadian Winter Sketch, in *Keoeeit—The Aurora* by W. Petrie.

469. Traprock, W. E., Ode to the Aurora, in *My Northern Exposure,* p. 197, Putnam's Sons, New York, 1922.

470. Tromholt, S., On the period of the aurora borealis, *Dan. Meteorol. Yearb., 1880,* 1, 1882.

471. Tromholt, S., Norwegian testimony of the aurora sounds, *Nature, 32,* 499, 1885.

472. *Tromholt, S., *Under the Rays of the Aurora Borealis: In the Land of the Lapps and Kvaens,* edited by Carl Siewers, Houghton-Mifflin, Boston, 1885.

473. Tromholt, S., *Catalog Nordlichter,* Jacob Dybwad, Oslo, 1902.

474. Trouvelot, E. L., *The Trouvelot Astronomical Drawings Manual,* Charles Scribner's Sons, New York, 1882.

475. Turner, L. M., Ethonology of the Ungava district, Hudson Bay Territory, *Bur. Amer. Ethnol. Annu. Rep. 1889-1890, 11,* 266, 1894.

476. Ungava, Publications of the Geological Survey of Canada, Anthropological Series *14,* 153, 19??.

477. Unterweger, J., Beiträge zur Erklärung der cosmisch-terrestrischen Erscheinung; uber das Polarlicht, *Dan. Akad. Wiss., 36,* 193, 1885.

478. *Vallance Jones, A., *Aurora,* D. Reidel, Hingham, Mass., 1974.

479. *Van Allen, J. A., Space and the radiation belt, *Time, 73*(18), 64, 1959.

Van Allen, J. A., Radiation belts around the earth, *Sci. Amer., 200,* 39, 1959.

480. Van Allen, J. A., The first public lecture on the discovery of the geomagnetically trapped radiation, report, State Univ. Iowa, 1960.

481. Vaussenat, M. C. X., Observatoire du Pic du Midi, *Ann. Soc. Meteorol. France, 34,* 118, 1886.

482. Vegard, L., On spectra of the aurora borealis, *Phys. Z., 14,* 677, 1913.

483. Vegard, L., Hydrogen showers in the auroral region, *Nature, 144,* 1089, 1939.

484. Vegard, L., Remark on Doppler-shifted hydrogen lines, *Trans. Ass. Terr. Magn. Elec., 484,* 490-491, Aug. 1948.

485. Virgil, *Aeneid,* Book 8, lines 524-529; *Georgics,* Book 1, lines 487, 492, ~25 B.C.

486. Volta, A., L'aurora boréale, in *Collezione dell'opere del Cavaliere Conte Alessandro Volta, Lettera Sopra l'aurora Boreale,* vol. 1, part 2, p. 425, G. Piatti, Florence, 1816.

487. Wace and Layamon, *Arthurian Chronicles,* Aldine Press, London, 1962.

488. Wagner, R., *The Ring of the Nibelung,* Act 4, *The Twilight of the Gods,* Kiesling, Leipzig, 1853.

489. Wakeman, C., Reflections on the morning of Christmas Day, 1819, North Georgia, in *The North Georgia Gazette and Winter Chronicle,* edited by W. E. Parry, p. 53, Murray, London, 1821.

490. Wang, P. K., and G. L. Siscoe, Some early descriptions of auroras in China, report, Univ. of Calif., Los Angeles, May 1979.

491. Wegener, A., *Thermodynamik der Atmosphäre,* J. A. Barth, Leipzig, 1911.

492. Weinstein, B., Aurora borealis, *Himmel Erde, 1,* 234, 360, 1889.

493. Weisskopf, V., *Knowledge and Wonder; The Natural World as Man Knows It,* Doubleday, New York, 1962.

494. Weyer, E. M., *The Eskimos, Their Environment and Folkways,* p. 243, Archon, Hamden, Conn., 1969.

495. Weyprecht, K., *New Lands Within the Arctic Circle,* vol. 1, by J. R. Payer, p. 328, 1940.

Weyprecht, K., and J. R. Payer, *Austrian Arctic Expedition 1882-1884.*

496. White, F. W. G., and M. Geddes, The Antarctic zone of maximum auroral frequency, *Terr. Magn. Atmos. Elec., 44*, 367, 1939.

497. White, R. S., *Space Physics*, Gordon and Breach, New York, 1970.

498. Whitman, W., *Complete Poetry and Selected Prose and Letters*, Random House, New York, 1938.

499. Whittier, J. G., *The Complete Poetical Works of John Greenleaf Whittier*, Houghton-Mifflin, Boston, 1894.

500. Whymper, F., *Travels and Adventure in the Territory of Alaska*, J. Murray, London, 1869.

501. Wilcke, J. C., Von der jährlichen und täglichen bewegungen der magnetnadel in Stockholm, *Sv. Vetensk. Acad. Handl., 38*, 273, 1777.

502. Wilcox, C., Sights and Sounds of the Night, in *Lyric America*, edited by A. Kreymborg, Coward McMann, New York, 1930.

503. Wilkes, C., *Narrative of the U.S. Exploring Expedition 1838–1842*, vol. 2, Lea and Blanchard, Philadelphia, 1845.

504. Williams, D. J., A case for magnetospheric research, *Eos Trans. AGU, 56*, 211, 1975.

505. Wilson, E. A., Illustrations of aurorae, in *Albumn of Photographs and Sketches from National Antarctic Expedition, 1901–1904*, Royal Society, London, 1908.

506. Wilson, E. A., *Diary of the Discovery Expedition to the Antarctic Regions 1901–1914*, Humanities Press, New York, 1967.

507. Winchester, J., Nature's fireworks show—The northern lights, *Readers Dig.*, 1978. (Various international editions for January 1978.)

508. Winckler, J. R., and L. Peterson, Large auroral effect on cosmic ray detectors observed at 8 g/cm^2 atmospheric depth, *Phys. Rev., 108*, 903, 1957.

509. Winn, J. S., Remarks on the aurora borealis, *Phil. Trans. Roy. Soc., 64*, 128, 1774.

510. Winthrop, J., *The History of New England from 1630 to 1649*, vol. 2, edited by J. Savage, Little, Brown and Company, Boston, 1853.

511. Wolf, C., *Gedanken über das ungewöhnliche Phaenomenon*, University of Halle, Halle, 1716.

512. Wolf, R., Astronomische mittheilungen, *Vierteljahreschri. Naturforsch. Ges. Zurich, 1–10*, 79, 1856–1859, *21–30*, 239, 1866–1872, *41–50*, 165, 1876–1879, *61–70*, 1884–1887.

513. Wolf, J. H., Jupiter, *Sci. Amer. 233*, 119, 1975.

514. Wolfius, or Wolf, C., *Reflections on the Works of Nature*, sect., 335, Halle, ↻ 1745.

515. Wordsworth, W., *The Complete Poetical Works*, Houghton-Mifflin, Boston, 1932.

516. Wrangel, F. P., *Narrative of an Expedition to the Polar Sea in the Years 1820, 1821, 1822, and 1823*, edited by E. Sabine, J. Madden Company, London, 1840.

517. Wright, C. S., Observations on the Aurora, in *British "Terra Nova" Antarctic Expedition 1910–1913*, Harrison and Sons, London, 1921.

518. Young, E., *The Complaint, or Night Thoughts (Night Ninth)*, Huntington, New York, 1813.

519. *Young, T., *Courses of Lectures on Natural Philosophy and the Mechanical Arts*, vol. 1, pp. 687, 716; vol. 2, p. 488, Johnson, London, 1807.

520. Zeyfuss, in *Aurora—Their Characters and Spectra* by J. R. Capron, p. 64, Spon, London, 1879.

521. Zölner, F., On the spectrum of the northern lights (in German), *Annden Phys. Chem., 141*, 574, 1870.

References added in press.

522. Bernhard, R., *The Hollow Earth*, Health Research, Mokelumne Hill, Calif., 1963.

523. *Brekke, A., and A. Egeland, *Nordlyset* (in Norwegian), Grondahl and Son, Oslo, Norway, 1979.

524. Brodribb, W. J., and W. Besant, *Constantinople*, p. 17, Seeley, Jackson and Halliday, London, 1879.

525. *Cincinnati Chronicle*, New celestrial phenomena, March 22, 1843.

526. Fisk, D. M., *Exploring the Upper Atmosphere*, chap. 7, Aurora, Faber and Faber, London, 1933.

527. Gardner, M. B., *A Journey to the Earth's Interior, or Have the Poles Really been Discovered*, chap. 14, The Aurora, E. Smith and Co., Aurora, Ill., 1920.

528. Guillemin, A., *Electricity and Magnetism*, chap. 5, Polar Auroras, revised and edited by S. P. Thompson, Macmillan, London, 1891.

529. Handie, E. M., *The Cry in the Midnight, or See the Bridegroom Coming*, Stockholm, 1883.

530. Knox, T. W., *The Voyage of the "Vivian" to the North Pole and Beyond*, Harper Brothers, New York, 1885.

531. Mitterling, P. I., *America in the Antarctic to 1840*, chap. 5, University of Illinois Press, Urbana, 1959.

532. Reed, W., *The Phantom of the Poles*, chap. 7, The Aurora, W. S. Rockey Co., New York, 1906.

533. Schröder, W., Zur Geschichte der Polarlichtforschung, *Physik. Blätter, 160*, 164, 1979.

534. Schröder, W., Some aspects of the history of auroral research, *Eos Trans. AGU, 60*, 1035, 1979.

535. Thomsen, I. L., Aurora Australis, in *The Antarctic Today*, chap. 12, edited by F. A. Simpson, A. H. and A. W. Reed and New Zealand Antarctic Society, Wellington, N.Z., 1952.

536. Weidler, I. F., *Commentatio de Aurore Boreali*, Wittenberg, 1739.

Films About the Aurora

1. "The Flaming Sky—Aurora." Produced by National Academy of Sciences, Washington, D. C. (16 mm color, sound, 27 minutes, 1960). Out of print. (General interest; history, mythology, current research, animation, auroral footage.)
2. "Color Television of Auroras." Produced by T. J. Hallinan and T. N. Davis (16 mm color, no sound track, 10 minutes, 1972). Available from Geophysical Institute, University of Alaska, Fairbanks, Alaska, 99701. (Technical; shows different types of auroral structures and movements.)
3. "Spirits of the Polar Night—The Aurora." Produced by R. H. Eather (16 mm color, sound, 30 minutes, 1973). Available from Film Center Associates, 992 Memorial Drive, Cambridge, Massachusetts 02140. (General interest; history, mythology, current research, interviews, auroral footage.)
4. "Discovery—The Aurora." Produced by Yorkshire Television, Leeds, England (16 mm color, sound, 27 minutes, 1975). (General interest; history, current research, interviews, animation, auroral footage.)
5. "Lights in the Northern Sky." Produced by Patrick Moore and T. Neil Davis (16 mm color, sound, 33 minutes, 1975). Available from Geophysical Institute, University of Alaska, Fairbanks, Alaska 99701. (General interest; mainly current research and television auroral footage.)
6. "Earthspace—The Magnetosphere." Produced by R. H. Eather and B. Kaufman (16 mm color, sound, 50 minutes, 1977). Available from National Audiovisual Service, Washington, D. C. 20409. (General interest; history, mythology, current research, interviews, animation, auroral footage.)
7. "Earthspace—Our Environment." Produced by B. Kaufman and NASA (16 mm color, sound, 16 minutes, 1978). Available from NASA Distribution outlets. (Shortened version of the film "Earthspace—The Magnetosphere.")
8. "Revontulet." Produced by Finnish Television, Helsinki, Finland (16 mm color, sound, 34 minutes, 1978). (General interest; history, mythology, current research, interviews, auroral footage.)

Name Index

Contains names specifically mentioned in text. See reference list for additional names. Page numbers in italics indicate photographs. Additional photographs of modern scientists are on pages 283–288.

A

Abney, F., 265
Abraham, 34
Aeschylus, 36
Ahmed ibn-Fadlan, 66
Aiken, C., 208, *208*
Akasofu, S.-I., 175, *175, 179*, 269, *269*
Akenside, M., 75
Alfvén, H., 218, *219*
Amanuensis, 63
Ames, N., 93, 94
Amundsen, R., 159, *159*
Anaxagoras, 36
Anaximenes, 35
Angot, A., 64, 119, *119*
Angström, A. J., ix, 81, 124, 164, *164*
Arago, D. F. J., 66, *78*
Argelander, S., 119
Aristotle, ix, 38, *38*, 39, 40, 42, 76
Arrhénius, S., 124, 134, *134*
Arthur (King), 187
Attila, 41
Aytoun, W. E., 199, *199*

B

Babcock, H. D., 165
Baly, E. C. C., 164
Banks, J., 143, *143*
Barr, E., 260
Baschin, O., 253
Batorij, S., 67
Bauer, W., 256
Baumhauer, M. E. H., 78, 124–125
Beals, C. S., 154
Becquerel, H., 64, 130, *130*, 237
Bellingshausen, T., 144, *144*
Benediktsson, E., 198
Berkner, L., 182
Bernard, R., 116
Bertholen, A., 64
Beyer, A. P., 46, *46*, 154
Bierman, L., 217, *217*
Biot, J. B., 64, 78, *78*
Birkeland, K., 129, 130–134, *131, 132, 133*, 136, 164, 217, 237, 247
Birkeland, O., 137
Biscoe, J., 144
Bock, C., *104*
Botley, C. M., *40*
Boué, A., 125
Bradley, F., 99
Bradley, L., 64
Brendel, M., 252, 253, *253*
Briggs, J. M., 59
Brjus, J. V., 68
Bronowski, J., 231
Browning, R., 196–197, *197*
Buddha, 79
Burns, R., 155, 195, *196*
Bush, D., 163
Butler, H. R., *98*
Byrd, R. E., 116, 206, *206*
Byron, G., 195, *195*

C

Caesar, J., 38, 39, 187
Camden, 44, 46
Canton, J., 64, 233, *233*
Capron, J. R., 64, *167*, 207, 233, *233*, 234, *234*
Carlheim-Gyllensköld, V., 87, *87*
Carrington, R., 79, 81, 126, 128
Castro Medice, R., 48, *48*

Cavendish, H., ix, 60, *60*, 62
Celsius, A., ix, 60, *60*, 69
Chapman, S., 164, *164*, 169, 173, *173*, 217, *218*, 268
Chaucer, G., 187, *187*
Chew, S., 96
Church, F., *102*
Coates, 64
Coleridge, S. T., 190, *190*
Collins, W., 188, *188*
Colp, A., 207
Cook, J., 143, *143*, 144
Cranch, C. P., 198, *198*
Creighton, D., 261, 262
Creutzer, P., 42, 43, *43*
Crookes, W., 132
Crowder, W., *156*

D

Dalton, J., ix, 60, 61, *61*, 62, 63, 64, 78, 81, 119–120, *120*, 124, 125, 126, 155
Daniken, E., 35
Davidialuk, D., *109*
Day, J., 98, *98*
Deacock, W., 260
de la Rive, M. A., 64, 233–234, *235*
de Mairan (see Mairan)
Democritus, 189
Derham, W., 52–53
Derwentwater, Lord, 281
Desaguliers, M., 59
Descartes, R., ix, 49, *49*, 64
Devik, K., 132
Devik, O., 129, *273*
Dickenson, E., 202, *202*
Dietrichsen, W., 132
Dionysius, 36, 38
D.J.A.M.R.D.C., 55
Dortous de Mairan (see Mairan)
Dostoyevsky, F. M., 196

E

Easley, J. A., 158
Eather, R. H., 260–277, *263, 264, 266, 267, 270, 271, 273, 278, 279*
Eddy, J. A., 125
Edlund, E., 64, 85, *85*
Egede, H., 107
Eigen, M., 64
Einstein, A., vi, 76
Eliot, T. S., 231
Episcopus, I., 41
Equalaq, *107*
Ernemann, 255
Estanda, 111
Euler, L., ix, 55, *55*, 59, 64, 70, 97, 116
Ezekiel, 33, 35, 37

F

Farquharson, J., 62
Fastie, W. G., 269
Feldstein, Y. I., 175, *175*
Felka, 70
Ferraro, V. C. A., 217, *218*
Foran, J. K., 201, *201*
Foster, J. R., 143, *143*
Francois, M., 48
Franklin, B., *viii*, ix, 64, 70, 94, 96–97, *97*, 98
Franklin, J., 76, *76*, 118
Frederick III, 49
Freya, 114
Fritz, H., 36, 76, *76*, 77, 78, 118–119, 122

Frobesius, J. N., 36
Früeauff, N. D., 50
Fulke, W., 44, *44*

G

Gabor, D., 243
Galileo, G., ix, 7, 48, 49, 51, *51*
Gardner, M. B., 116
Garriott, O. K., 227
Gartlein, C. W., 158, 169, *174, 175*, 255
Gassendi, P., ix, 48–49, *49*, 51
Gauss, C. F., ix
Gay Lussac, J. L., *78*
Geinzius, G., 69, *69*
Gemma, C., 42, *43*
Gervasius, 42
Gilbert, W., 215, *217*
Gmelin, J. G., 159
Goethe, W., *190*, 195
Gold, T., 217, *217*
Goldstein, E., 130, 237
Gough, J., 155
Greely, A. W., 62, *62*
Greenwood, I., 96, *96*
Gregory of Tours, 41, *42*, 51
Guidicci, M., 51

H

Haavio, M., 106
Halley, E., ix, 50, 51, 52, *52*, 55, 59, 60, 62, 63, 64, 95, 96, 97, 116, 188
Haliburton, J. C., 207
Hamilton, S., *72*, 127, *204*
Hannibal, 186
Hansteen, C., 119, *119*
Häuy, M. R. J., 63
Hawksbee, F., 64, 233
H. C. B., 210
Hearne, S., 111, 155, *155*, 190
He-Holds-The-Earth, 111
Heikkila, W., *230*
Heinsius (see Geinzius)
Heitman, J., 55
Hellman, G., 43
Henkel, 70
Herrick, E. C., 99, *99*
Hesiod, 36
Heyl, P. R., 245
Hiawatha, 111
Hiorter, O. P., 60, 69
Hippocrates, 36
Hitler, A., 106
Hodgson, R., 81
Holden, A. P., 64
Holyoke, E. A., 96, *96*
Homer, v
Hopkins, G. M., 206–207, *207*
Hughes, K. J., 260
Hulburt, E. O., 169, *169*
Humboldt, A., ix, 64, 161, *161*

I

Italicus, S., 186
Ithenhiela, 111

J

Jacka, F., 260
Jacob, V., 199
Jaikema, F., *73*
Jeremiah, 35, *35*
Jesse, O., 253

J. F. R., *53-54*
Job, 117
Johannsen, 159
Jones, J., 50
Justin II, 79

K

Kämtz (Kaemtz), L. F., 76
Kaufman, B., *276*
Keats, J., 195, *195*
Keimatsu, M., *40*
Kenmure, Lord, *281*
Kennedy, J. F., 263
Kepler, J., ix, 48
Kirill (Monk), 67, *67*
Kirwan, R., 64
Knox, T. W., 242
Kochen, 68
Kraft, W. G., 69, *69*

L

Layamon, 187
Lemström, S., 64, 233-234, *233, 235*, 236, *236*
Lewis, R., 96
Libes, M., 63, 64
Lindemann, F. A., 164, *164*
Link, F., 36
Livy, 38
Lomonosov, M. V., 65, 70-73, *70, 71*, 189
Longfellow, H. W., 111, 114, 185, *186*
Loomis, E., 62, 76, *76, 77*, 94, 99, *99*, 126
Lord, J. V., *208, 209*
Lovering, J., 99, *99*
Lucanas, M., 187
Lucretius., 189
Ludwig, G., *181*
Lycosthenes, C., 35, *35*, 36, 39, *39*, 43

M

Magnus, G., 41
Mairan, J. J. D., 51, 55, *55, 56, 57*, 59, 60, 62, 64, 67, 69, 96, 97, 124, 126, 143, 147, 189
Maraldi, M., 53
Marensson, J. A., 55
Marsh, B. V., 64
Marston, G., *142*
Masterman, J., 267
Mather, C., 94, *94*
Mathew of Westminster, 41
Mathieu of Edesse, 41
Maunder, E. W., 49, *50*
Mawson, D., 145, *147*
Mayer (see Meyer)
McAdie, A., 122, *122, 123*
McIlwain, C. E., 173, 181, *181*
McLennan, J. C., 165, *165*
Medice, R. C. (see Castro Medice)
Meinel, A. B., 169
Mende, S. B., 268, 270, *271, 272*, 274, *274, 275, 276, 277*
Menshutkin, 70
Mercredi, 110
Meredith, C. H., 173
Meyer, F. C., 69
Milman, H. H., 190, *190*
Milton, J., 47, 244
Mion, P., 228-229
Mohammed, 79
Moltke, H., *29, 191-194*
Monge, M., 63, 64
Mozer, F., 265, *265*
Muncke, G. W., 76
Musschenbroek, P., 61, *61, 64*

N

Naainas, 115
Nadubovich, Y. A., 73
Nanahboozho, 111
Nansen, F., *88, 89*, 107, 205, *205*
Neubrigensis, G., 42

Noceti, C., 189, *189*
Nordland, O., 107

O

O'Brien, B. J., 263, 264, 265, 267
Obsequens, J., 36
Odin, 114
Ogilvie, W., 157
Olmsted, D., 64, 78, 81, 94, 99, *99*

P

Palmer, J. C., 141, 144
Parrot, G. F., 64
Parry, W. E., 62, 198
Pastore, J., 247
Paulsen, A., 130, *130*, 169
Payer, J., *90*
Peter the Great, 68, *68*
Petrie, W., 255
Philip of Macedonia, 36
Pius XI, 106
Planté, G., 64, 234, *235*
Pliny the Elder, 38, *38*
Plutarch, 36
Podgorny, I. M., *237*, 239
Ponting, H. G., 145, *145*
Pontoppidan, E., 59-60, *59*, 97
Proctor, R. A., 81, *81*
Puskin, A. S., 196

Q

Quetelet, A., 124

R

Radowitz, Baron von, 92
Ramsey, W., 164
Rasmussen, K., 110
Reed, W., 116
Reyes de Castro (see Castro Medice)
Rhoden, A., 49
Richardson, J., 118
Robert, C. G. D., 155
Roberts, W. O., 122
Robertson, T. A., 155
Robie, T., 95
Roederer, J. G., 217
Ross, J. C., 198
Rowell, G. A., 64
Rust, C. R., 161

S

Savage, R., 188
Schove, D. J., 36, *40*
Schuster, A., 134, *134*, 164
Scott, R. F., 103, 144-145, *145*
Scott, W., 103, 195, *195*
Sedden, Rev., 128
Seneca, ix, 39, *39*, 42, 92
Service, R., vi, 155, 185, 210-213, *210*
Sewall, S., 95, *95*
Sguario, E., *58*
Shackleton, E. H., *143*, 145, *147, 151*
Shakespeare, W., 47, 187
Shrum, G. M., 165
Silberman, M. S.-J., 64
Silverman, S., 154, 155, 157, 159
Singer, S. F., 184
Siple, P., *278*
Siscoe, G. L., 51
Smart, C., 189, *189*
Smyth, P., 119, *119*, 124
Solzhenitsyn, A. I., 196
Southey, R., 195, *196*
Spence, J., *107*
Starck, C. H., 50
Stevens, W., 163, 202-203, *202*
Stiles, E., 98, *98*
Stockmar, Baron, 92
Stokes, F. W., 196
Störmer, C., 2, 62, 134-136, *135, 136, 137, 138, 139*, 145, 184, 254-256, *255*

Stothers, R., 36, 38, 39
Sullivan, W., 182
Symmes, J. C., 116
Synge, J. M. 204
Szent-Gyorgi, A., 259

T

Tacitus, P. C., 154
Tatiscev, V. N., 66, 68
Taylor, B., 206, *206*
Taylor, B. F., 199, *199*
Tennyson, A., 195, *195*
Thomas á Becket, 42
Thomson, J., 187-188, *187*
Thomson, J. J., 132
Tiberius, 39
Toll', E. V., 73, *73*
Traprock, W. E., *203*
Tromholt, S., 84, *84*, 85, 86, 154, 236, 253, *254*
Trouvelot, E. L., *100-101*
Trumbull, B., 95
Tuan, T. F., 154, 155, 157, 159

U

Ulff, C., 154
Ulloa, D., 143, *143*
Unterweger, J., 64, 91-92
Urbano, P., 45

V

Van Allen, J., 171, 173, *180*, 181, *181*
Vaussenat, C. X., 236, *237*
Veeder, M. A., 128
Vegard, L., 165, *168*, 169
Vespasian, T., 39
Virgil, 186, 259
Volta, A., ix, 64, 78, *78*
von Braun, W., 178
Vrangel (see Wrangel)
Vysehrad, C., 41

W

Wagner, R., 204
Wakeham, C., 198
Weber, E. J., *271, 272, 273*
Wegener, L., 164, *164*
Weinstein, B., 253
Weisskopf, V., 75
Westphal, D., *273, 276*
Weyprecht, K., *90*, 118, 205, *205*
White, R., 260, 262
Whitman, W., 155, *155*
Whittier, J. G., 200, *200*
Whymper, F., *160*
Wilcke, J. C., 52
Wilcox, C., 195
Wilkes, C., 144, *144*
Williams, D. J., 243
Wilson, E. A., 144, *145, 146*, 251, 253
Wilson, R., 247
Winn, J. S., 119
Winthrop, J., 96, *96*
Winthrop, J. (Gov.), 95, *95*
Wolf, C., 59, 62, 70
Wolf, R., 126
Wolfius (see Wolf, C.)
Wordsworth, W., 153, 155, *155*
Wrangel, F. P., *72*, 73

X-Z

Xenophanes, 35
Young, E., 1
Young, T., 63, *63*
Zechariah, 35
Zölner, F., 81

Subject Index

Page numbers in italics indicate illustrations.

A

aborigines (see Australian mythology)
acceleration mechanism, 3, 224, 245
acies igneae, 41
air flows (see atmospheric air movements)
aircraft studies, 147, *152*, 267–270, *267, 269*
airglow, 4, *4, 248*, 276
alignment of aurora, 59, 85, 119
alliteration (poem), 210
all-sky
  photographs, *4, 16, 17, 20, 21, 37, 176*
  paintings and drawings, *57, 92, 192*
all-sky cameras (see cameras)
alpha particles, 265
altitude (see height)
amateur observations (IGY), 173, *174*
angstrom, 164
Antarctica
  early expeditions, 144
  U.S. program, 272–273
Arab traveler's description, 66
arcs (see forms)
Arctic expeditions, 76, 84, 87
Argus tests, 184
artificial aurora, 231–241, *236, 237, 238, 239, 240*
  Argus tests, 184
  de la Rive, 233, *235*
  Lemström, 233, *235, 236*
  Planté, 234, *235*
  rocket electron accelerators, *240*, 241, 245
  space shuttle, 241, *246*
Aristotle's theory, 38
astronauts, *227*
atlas of auroral forms, 2
atmosphere (earth's), 165, 175, 217
atmospheric air movements, *viii, ix*, 55
ATS (see satellites)
audibility (see sounds)
aurora (other planets), 248–250, *250*
Aurora (Roman goddess), v
aurora australis (see also under specific topics), 49, 79,
    141–152
  drawings, *142, 146, 148–149, 150, 151, 152*
  early observations, 143–144
  origin of name, 49
  presence suspected, 143
aurora borealis (see under specific topics)
  origin of name, 49, 51
Aurora Data Center, 173
auroral oval (see oval)
auroral zone (see zone)
Australian mythology (see folklore)
Australian National Antarctic Research Expedition, 260
author's personal involvement, 260–277

B

Ballad of the Northern Lights, vi, 212–213
balloon flights, 170
bands (see forms)
band spectrum, 165, *169*
barium clouds, *232*, 245, *245, 246*
barometric changes, 122, *123*
belts of maximum frequency (see probability
  and zone)
biblical references, 33, 34–35, *35*, 38, 66, 79, 106, 117
black aurora, *20*
blood and aurora, 41, 42, 44, *45*, 79, 104
bombardment (see precipitating particles)
Boston College, 272
bounce motion (see mirroring)
boundary layer, 218, *230*
breakup (see substorm)
bremsstrahlung radiation, 170
brightness (see intensity)

brightness classification, 2, 257
brush discharge, 159
Buddha's lights, 79
Byzantium, 36

C

cameras
  all-sky, 173, *174*, 175, *176*, 179, 226
  settings for auroral photography, 257, 258
  Störmer's, *137*, 138, 255, *255, 256*
cartoons, *162, 250*
catalogs
  auroral sounds, 155
  visual aurora, 36, 66, 67
cathode rays, 130, 132, 134, 237
cause of aurora (see theories)
cave drawings, 34
Ceylonese mythology (see folklore)
Chapman-Ferraro cavit, 217, *218*
Chapman publications, *173*
charged particles (see also protons and electrons)
  motions, 132, *135, 136*
  trapping, 135, *136*, 184
chasmata, 38, 39, 40, 67
children's poem, 208, *209*
Chile (auroral reports), 143
Chinese reports, 40, *41*
chronicles
  Anglo-Saxon, 41
  Irish, 41
  Russian, 66–68
  Scottish, 41
Churchill, 264, *264*, 265, 267, 268, 273, 274
cinematography (see films)
circumpolar cultures, 107
cirrus clouds (see clouds)
classic poems, 186, 187
classification system, 2
cleft (see polar cusp)
climate, 122, *124*, 125
clouds, 118, *118*, 119, 130
coastline effect, 73
colonial America, 93, 101
color photographs, 9–29
  first, 255, 256
  recommendations, 256–258
colors of aurora, 3, 165, 220
color threshold of the eye, 165
comets, 51, 164, 186, 217, *218*
communications (auroral effects on), 247
conjugacy, 78, 147, *152*
constant time flights, 268
continuous spectra, 81, 164, *169*
corona (see forms)
corpuscular radiation, 130, 147, 164
cosmic origin of aurora, 78, 85, 91
cosmic rays, 135, 181
Cro-Magnon times, 34
currents, 92, 130, 145
  field aligned, 130, *220*, 221
  ionospheric, 145, 221
  magnetospheric, *220*, 221, 226
curtains (see drapery)
cusp (see polar cusp)
cycles (see periodicity)

D

DAPP (see DMSP)
Dark Ages, 41
dayside aurora, 147, *177, 179*, 220, 269
  detection on Jupiter, 249
dayside cleft (see cusp)
dayside cusp (see cusp)
death (auroral links), 104, 107, *107*, 115, 196

depletion regions, 4, *4*
descriptions of aurora, 60, 144, 161, 185, 205–207, 261
diffuse aurora, 173, 221
discharge tube, 60, 233, 234, *234*
discrete spectrum (see line spectrum)
dissertations (first), 50
divine justification, 67, *67*
documentary films (see also reference list), 273–274, *273,*
    275–276, 276
doppler shift, 169, 170
DMSP (see satellites)
dragons, *37, 40, 40,* 41, *44, 45,* 187
drapery (auroral form), 2
drawings (see also first drawings)
    additional drawings, *79, 82, 83, 86, 87, 88, 89, 90, 91,*
    92, 112, 113, 118, 120, 121, 131, 148, 149, 150, 157, 158,
    160, 190
dynamo action, *218*

E

earliest reports
    China, 40
    Greece, 35
    England, 41
    New England, 95
    Old Testament, 34, 35
    Roman, 38
    Russian, 66
    southern hemisphere, 143, 144
earthquake lights, 128
earthquakes, 52, 53, 125, 128
Earthspace (film), 275–276, *276*
electric discharge, 60, 64, 85, 218, 233, 234, 237
electric fields, 64, 159, 218, 221, 228, 245
electric nature of aurora, ix, 59, 64, 84, 85, 119, 219, 233
electrical experiments, 59, 60, 70
electromagnetic gun, 134
electromagnetic furnace, 134
electron accelerator, *240,* 241
electron aurora, 173
electrons, 132, 134, 173, 184, 217, 223, 237, 241
electrostatic repulsion, 134, 164
eleven-year period (see periodicity)
Elizabethan England, 44
energy in auroras, 226, 241, 242
Enlightenment, 59
entry layer, 218, *220,* 230
Eos (Greek goddess), *v*
epic poems, 186, 187
equatorial airglow, 4, *4*
equatorial aurora (see great auroras)
eskimos, 158
    auroral sound, 110, 158, 161
    beliefs, 107, *107,* 110
    sculpture, *109*
Explorer satellites
    I, 178, 181, *183,* 228
    II, 181
    III, 181, 184
    XIV, 217
Ezekiel, 33, 35, 57, 66

F

Fatima, 106
fear of aurora, 38, 41, 44, 95, 111, 112, 113
ferruginous particles, 61, 63, 78, 81
field-aligned currents (see currents)
films of aurora (see also reference list), 255–256, 273,
    275
fine structure, 245
Finnish folklore (see folklore)
fires in sky (see heavenly fires)
first—
    association with magnetic disturbances, 60, 69
    auroral oval concept, 87
    auroral zone concept, 76
    description (see earliest reports)
    dissertations, 50
    drawings, *35, 39, 40, 41, 43, 45, 46, 53, 54, 57, 58, 71, 74*
    films, 255–256
    height measurements, 60, 61, *61,* 62, 69, 130

photographs, 253
    aurora australis, 145, 147
    aurora borealis, *252,* 253–254
    color, 254–255
    films, 255–256
poetic references
    classical, 186
    English, 187
printed description, 43, *43*
reports of sounds, 154
scientific illustration, *43*
sighting of aurora australis, 143
textbook, 55, *56*
theories
    American, *viii,* ix
    English, 44, 50
    Greek, 38
    Russian, 69, 70
    Scandinavian, 42
use of term aurora borealis, 49, 51
woodcuts, *35, 39, 45, 46*
First Polar Year, *82–83,* 84, 87, *191–194,* 253
flaming aurora, 2
flares (see sun)
flickering (see pulsating)
folklore, 103–115
    Australian, 143
    Ceylonese, 79
    Eskimo, 107, 110
    Estonian, 106
    Finnish, *105,* 106
    Greenland, 107, 110
    Indian, 111
    Lapp, 112, *112,* 113, *113,* 115
    Latvian, 114
    New Zealand, 143
    Norwegian, 104, 114
    Russian, 112, 114, 115
    Scottish, 114, 188
football (spirits playing), 107
forecasting (see prediction)
forgotten explanation (see humor)
forms
    arcs, 2, *8–9, 11, 16, 20, 22, 30*
    bands, 2, *10–14, 16–23, 25–29*
    corona, 2, *10, 11, 15, 21*
    patches, 2, *20*
    rays, 2, *10–14, 19, 25–27, 29*
    veils, 2, *27*
Fort Churchill (see Churchill)
Fred's electric nursery (see humor)
freezing breath, 159
frequency of aurora (see periodicity)
future experiments, 245

G

gales (see winds)
Galileo-NASA research plane, 267, 268, 269, 270
gegenschein, 4
geiger counters, 181, *181,* 184
Genesis, 34
geocoronium (see spectrum)
geographic distribution, 3, *4,* 76, *77*
geomagnetic latitude, 79, *79,* 147
geomagnetic pole, 76
Gillam, 272, 275
great auroras, 79, *79,* 143, *150*
Greek literature, 35–36
green line, 164, 165, *169*
Greenland folklore (see folklore)

H

Harvard, 95, 96, 99
heavenly battles (see military association)
heavenly fires, 35, 36, 39, 40, 41, 43, 44, 46, 106, 111
height, 52, 60, *61,* 62, 98, 122, 130, 136, 165, 168
    change of color with, 3, 165, 220
    first measurements, 60–62, 69, 130
    Störmer's measurements, 136, 138
    triangulation, 52, 60, 61, *61,* 62, 136
helium emissions (see spectrum)

320

high-altitude nuclear tests (see Argus)
history
  American, 93–101
  biblical, 34, 35, *35*, 38
  Chinese, 40, *41*
  English, 41, 42, 44, 46
  French, 41, 46
  Greek, 35, 36, 37
  Roman, 37, 38
  Russian, 66, 69
  Scandinavian, 42, 46
hollow earth theories (see theories)
homogeneous forms (see structure)
humor
  cartoons, *162*
  forgotten explanation, 126
  Fred's electric nursery, 242, *242*
  greeting card, *162*
  mistaken aurora, 92
  poetry, 199, 203, *203*, 208, 209, 210, 213
hydrogen aurora (see proton aurora)
hydrogen beams, 130, 169, 237
hydrogen emission (see spectrum)
hydrogeneous gases, 63

I

I.B.C., 257
ice particles (see reflections)
image intensifier systems, 4, 256, 270, *274*, *275*
infrared emissions, 226
Indian beliefs
  North American, 111, 113
  Siberian, 113
instabilities, 221, 245
intensity, 2, 85, 257
intensity indices (see I.B.C.)
international brightness coefficient (see I.B.C.),
international cooperative programs
  International Geophysical Year, 147, 170, 171–184
  International Magnetospheric Study, 276
  First Polar Year, 84, 87, 191–194, 253
  Second Polar Year, 73
interplanetary magnetic field, 218, 224, 245
ionospheric currents (see currents)
ionospheric scintillations, 247
Isis (see satellites)
isochasms, 76, 119

J

Jeremiah, 35, *35*
Jerusalem (auroral appearance), 39
Job, 117
Johnson Island tests, 239–241, *239*
Jupiter, 124, 249, 250, *250*

K

King Arthur's dream, 187
King's Mirror, 42, *154*
krypton spectral line (see spectrum)

L

land of the northern lights, 111
Lapp folklore (see folklore)
Latvian folklore (see folklore)
legends (see folklore)
lifetime (see metastable)
light to hunt or travel by, 85, *112*, *113*
lightning (see thunderstorms)
line spectrum, 81, 164, *169*
literature (aurora in), 205–207
Lockheed Research Laboratories, 268
Lord Derwentwater story, 281
low auroras, 62, 130, 236
low-latitude aurora (see great auroras and
  mid-latitude aurora)

M

macaronis, 34
Maccabees, 38, 190
magnetic cusp (see polar cusp)

magnetic disturbances (see magnetic nature)
magnetic field
  guiding particles, 3, 134, 135, *135*, *136*, 169, 170, 184, *184*,
    241
  magnetic poles, 76, 85, 130
  reversals, 34
  wandering of poles, 34
magnetic mirror point (see mirror point)
magnetic nature of aurora, 52, 60, 61, 69, 76, 84, 130
magnetic storms, 4, 30, 31, 81, 122, 164
magnetic zenith, 2, 52
magnetical effluvia, 50, 52, 63
magnetohydrodynamic dynamo (see dynamo)
magnetopause, *219*, *220*, *224*, 229, 230
magnetosheath, *219*, *225*, 230
magnetosphere, 173, 214, 217–230, *219*, *220*, *224*, *225*,
  228–229, 230
  first use of term, 217
  of other planets, galaxies, 249, 299
magnetospheric currents (see currents)
mantle, 218, *219*, *220*, 230
Maori legends (see New Zealand folklore)
map of auroral zone, *4*, 76, *77*
M-arcs (see red arcs)
Mariner (see satellites)
Mars, 145, 249
mathematical formalization, 63
Maunder minimum, 49, *49*, 51, 50, 52, 68, 94
Mawson (antarctic station), 260, 263
Mercury, 248
merging, 218, 224, *224*, 245
merry dancers (see names for aurora)
metastable transitions, 165
meteorological phenomena (see weather)
meteors, 61, 124, 128, 186
midday aurora (see dayside aurora)
Middle Ages, 41, 42, 66
mid-latitude aurora, 34, 96
military association, 36, 38, 39, 40, 41, *45*, 50, 66, 67, 69
military interest, *123*, 247
miracles, *48*, 106
mirror machines, 248
mirroring of particles, 135, *136*, 184, *184*
mistaken aurora, 39, 92
model experiments, 132, *132*, *133*, 233, 234, *235*,
  237–238, *237*, *238*, 239
monitoring and prediction, 5–7, *7*
moon, 124
motions of aurora, 2, *16–17*, *20*, 245, 253
movies of aurora (see films)
music, 186
mythology (see folklore)

N

names for aurora, 280–281
  first use of term aurora borealis, 49
  Russian, 69, 70
neon light tube analogy, 3
neutral point, 224, *224*
New England (first auroral reports), 94, *94*, 95
newspaper comments, 5, *183*
New Zealand folklore (see folklore)
nightglow (see airglow)
nitrogen emissions (see spectrum)
nitrous gases, 63, 76, 81
Nobel Prize (Alfvén), 218, *219*
noise of aurora (see sound)
Norse mythology (see also folklore), 114
northern lights (see also specific topics)
  title of poems, 198, 199
northern light people, 111
Norwegian mythology (see folklore)
Norwegian research, 129–136
nuclear explosions, 184, 239, *239*, 241

O

occurrence (see probability)
odor, 159, 161
OGO D (see satellites)
oil pipe line, 248, *249*
Old Testament, 33, 34, 35, *35*, 38, 66, 106, 117

Olympia myth, 36
omens, 38, 40, 48, 49, 52, 69, 94, 95, 97, 103, 113, 200
orientation (see alignment)
Ostia (auroral appearance), 39
oval (auroral), 87, 175, 177, 179, 221, 226
oxygen emissions (see spectrum)
ozone, 161

P

paintings,
  Howard Russell Butler, 98
  Carl Bock, 104
  Frederick Church, 102
  William Crowder, 156
  Stephen Hamilton, 72, 127, 204
  F. Jaikema, 73
  Harald Moltke, 29, 191-194
  Frank William Stokes, 196
  Carl Störmer, 139
  E. L. Trouvelot, 100, 101
  E. A. Wilson, 146
pamphlets
  first printed auroral description, 43, 43
particle mirroring (see mirroring)
particle precipitation (see precipitation)
patches (see forms)
periodicity
  eleven year, 7, 7, 80, 125, 126, 132
  secular, 55, 79, 99
  twenty-seven day, 5, 7, 125
perspective effects (see corona)
phosphorus compounds, 63, 64
photochemical reactions (see airglow)
photograph gallery of auroral scientists, 283-288
photographic atlas, 2
photography, 251-258
  attempts, 85, 145, 253
  first black and white, 252, 253-254
  first color, 254-255
  first film, 255-256
  modern filming, 256
  recommendations for photography, 256-258
photometers, 260, 264, 267, 270, 272, 275, 277
photometric imaging, 270, 274, 275
Pioneer (see satellites)
planets (aurora on others), 248-250, 250
plasma, 164, 243
plasma bubbles, 4
plasma instabilities (see instabilities)
plasma sheet, 218, 219, 220, 221, 224, 225, 229, 230, 245
plasmasphere, 218, 219, 220
poetic references, 1, 153, 155, 185-203, 207, 210-213,
    279
  first, 186
  first in England, 187
polar cleft (see polar cusp)
polar cusp, 218, 219, 220, 220, 225, 228, 230
polar explorers, 76
  sketches from expeditions, 88, 89, 90, 91
polarization of spectra, 78
politics and aurora, 53
portents (see omens)
postage stamps showing aurora, 140, 159
power (see energy)
power transmission lines, 247
precipitating particles, 4, 7, 169, 173
prediction of aurora
  newspaper warnings, 1, 5
  publications available, 5
  tape-recorded message, 5, 6
pressure (see barometric)
probability of aurora
  major cities, 3
  map, 4, 76, 77
prodigy (see omens)
prognostic (see omens)
prose descriptions (see descriptions)
protons, 169, 173, 260
proton aurora, 169, 173, 260, 263, 265
pulsars, 249
pulsating aurora, 2

Q

quenching, 165
questionaires on auroral sounds, 154
quiet aurora, 2

R

radar—radio, 247
radiant point (see corona)
radiation belts (see Van Allen)
radio stars, 249
rain, 119, 120
Rand report (on satellites), 175
rays (see forms)
rayed form (see structures)
red arcs, 4
red aurora, 24, 26, 27, 39
reflections (as explanation of aurora), 36, 48, 49, 55, 59,
    63, 68, 78, 84, 104
religious significance (see also Old Testament), 106
  Cincinnati meteor, 92, 92
  Russia, 66
Renaissance, 46, 48
Repu (Finnish legend), 105, 106
research papers, 217
reverence to aurora, 113
Rice University, 263, 264
Roberval, 275
rockets, 164, 170, 172, 173, 240, 241
Roman references, 38-39
Rouffignac cave drawings, 34
Russia, 65-73
  earliest auroral reports, 66
  first scientific work, 69-70
  folklore, 112
  words for aurora, 69

S

SAR-arcs (see red arcs)
satellites, 164, 175, 178, 181-182, 183, 184, 247
  early American problems, 175, 178
  ATS, 272, 275
  DMSP, 8, 30-31, 179, 216, 226
  Explorer I, II, III, XIV, 178, 181, 183, 184, 217, 228
  Isis, 226
  Mariner, 248, 249
  OGO D, 222
  Pioneer, 249
  Sputnik I, II, 175, 178, 182, 183, 184, 228
  Vanguard, 178
  Viking, 249
  Voyager, 249, 250
satellite data, 223
satellite photographs, 8-9, 30-31, 179, 216, 226, 227, 277
scaling of model experiments, 238
scientific explanations (see theories)
scintillations (see ionospheric scintillations)
Scottish folklore (see folklore)
sculpture (see Eskimo)
seasonal variation, 55
Second Polar Year, 73
secular variation (see periodicity)
serpents (see dragons)
shooting stars, 124
shoreline effect (see coastline)
shuttle (see space shuttle),
Siple, 275, 277
Skylab, 227
solar activity, 5, 122, 128
solar flare (see sun)
solar wind, 3, 5, 7, 217, 218, 219, 220, 220, 221, 224, 225,
    226, 228-229, 230
  entry mechanism, 218, 220, 221, 224, 245
sounds, 59, 110, 153, 161
  attempts to record, 155, 161
  earliest reference, 154
  effect on animals, 159
  Eskimos, 158, 161
  overhead aurora, 157
  poetic references, 155
  possible causes, 157, 159
  questionnaires, 154

South Atlantic anomaly,   265
South Pole,   *152, 272, 273, 275, 278*
southern hemisphere (first auroral reports),   143, 144
southern lights (see aurora australis)
Space Environment Service Center,   5, 6, 7, *7*
space shuttle,   241, *246*, 247
space research,   173, 175, 215–230, 241, 247, 248–250
spacecraft charging,   247
spectroscopy,   81, 124, 163–170, 260
spectrum,   81, 164, 166, 169
  comparison with other spectra,   124, 164, *166*
  first measurements,   81, 164, 166
  geocoronium,   164
  helium lines,   265
  hydrogen lines,   *166*, 169, *169*
  krypton lines,   164
  nitrogen lines,   165, *169*
  oxygen lines,   165, *169*
spirits of the aurora,   107, 109, 110, 111, 114, 115
Spirits of the Polar Night (film),   *12–13, 16–17*, 273–274, *273*
Sputnik satellites,   175, 178, 182, *183*, 184, 228
  announcement,   182, 183
  effect in United States,   175, 178
stamps (see postage stamps)
Stone Age auroras,   34
storms
  magnetic,   4, 30, 31, 122, 132, 176
  meteorologic,   118–119
striated forms (see structure)
structures,   2
  homogeneous,   2
  rayed,   2
  striated,   2
substorm,   175, 179, 221, 224, *224*, 226, 249
  trigger mechanism,   224, 245
subvisual aurora,   221, 268
sulphurous vapors,   53, 76
sun,   1, 7, *7*, 178, 225, *244*
  control of weather,   122, *124*, 125
  inner (see hollow earth)
  solar flares,   6, 7, *7*, 79, *80*, 81, 130, 225
  solar wind (see under solar wind)
  source of auroral particles,   130, 164, 237
sunspots,   7, 55, 76, *80*, 81, 95, 125, 132, *178*
  sunspot cycle,   7, *7*, 76, *80*, 132
supersonic airliners,   299
superstitions (see folklore)

T

tail of magnetosphere,   217, 221, 224
telegraph/telephone lines,   85, 247
telephone (see telegraph)
television systems (see TV)
temperature,   119
terella,   132, *132, 133*, 134, 237, *237, 238*, 239
terms for aurora (see names)
textbooks
  first,   55, *56, 57*
  second,   55
theater,   204
Theatrum European,   49

theories of aurora,   64
  pre-1900 (see under first theories)
  hollow earth,   116, *116*
  100 years ago,   64
  modern,   3, 217–218, 220–221, 224–228
thermonuclear fusion,   239, 248
thunderstorms,   104, 119, 122
tilting filter photometer,   260
tinnitus,   157
trajectories,   3, 132, 134, 135, *135, 136*, 184, *184*
trapped particles,   135, *136*, 184, *184*
triangulation (see height)
tropical auroras (see great auroras)
Turkey (see Byzantium)
TV set analogy,   226, 228
TV systems,   241, 247, 256, 270, *274*, 275
twenty-seven-day period (see periodicity)

U

ultraviolet light theory,   169
ultraviolet radiation,   130, 226, 249, 299
unanswered questions,   245, 249

V

Valkyrior,   114, 154
Van Allen radiation belts,   135, 181, *181*, 184, *184*, 224
Vanguard (see satellites)
veil (see forms)
Venus,   124, 249
vertical extent (see height)
Viking (see satellites)
Vikings,   42, 114, 249
visual observations,   87, *87*, 173, *174*
volcanoes,   69, 78
Voyager (see satellites)

W

walrus skull,   107, 110
warning of aurora (see prediction)
wars and famine (see military association
  and omens)
wave-particle interactions (see instabilities)
weather,   69, 104, 119–122, *123*, 124, *124*, 125, 130, 145, 248
where to see aurora,   3
whistling,   110, 113
wind,   119
woodcuts (see first woodcuts)
  Nansen,   *88, 205*

X–Z

X rays,   170, 226
X ray stars,   249
Yale University,   98
yellow-green line (see green line)
Zechariah,   35
zodiacal light,   55, 69, 124, 134, 164
zone (auroral),   76, 175
  northern,   76, *77*
  southern,   147